PENGUIN BOOKS

FOOLED BY RANDOMNESS

Nasim Nicholas Taleb has devoted his life to immersing himself in problems of luck, uncertainty, probability and knowledge. Part literary essayist, part empiricist, part no-nonsense mathematical trader, he is currently the Dean's Professor in the Sciences of Uncertainty at the University of Massachusetts at Amherst. *Fooled by Randomness* has been published in eighteen languages and was selected by *Fortune* magazine as one of 'The Smartest Books of All Time.' His most recent title is *The Black Swan: The Impact of the Highly Improbable*. Taleb lives (mostly) in New York.

FOOLED BY RANDOMNESS

•

The Hidden Role of Chance
in Life and in the Markets

SECOND EDITION, UPDATED BY THE AUTHOR

Nassim Nicholas Taleb

PENGUIN BOOKS

PENGUIN BOOKS

Published by the Penguin Group
Penguin Books Ltd, 80 Strand, London WC2R 0RL, England
Penguin Group (USA) Inc., 375 Hudson Street, New York, New York 10014, USA
Penguin Group (Canada), 90 Eglinton Avenue East, Suite 700, Toronto, Ontario, Canada M4P 2Y3
(a division of Pearson Penguin Canada Inc.)
Penguin Ireland, 25 St Stephen's Green, Dublin 2, Ireland
(a division of Penguin Books Ltd)
Penguin Group (Australia), 250 Camberwell Road, Camberwell, Victoria 3124, Australia
(a division of Pearson Australia Group Pty Ltd)
Penguin Books India Pvt Ltd, 11 Community Centre, Panchsheel Park, New Delhi – 110 017, India
Penguin Group (NZ), 67 Apollo Drive, Rosedale, North Shore 0632, New Zealand
(a division of Pearson New Zealand Ltd)
Penguin Books (South Africa) (Pty) Ltd, 24 Sturdee Avenue, Rosebank, Johannesburg 2196, South Africa

Penguin Books Ltd, Registered Offices: 80 Strand, London WC2R 0RL, England

www.penguin.com

First published in the United States of America by Texere 2004
This edition first published by The Random House Publishing Group 2005
Published in Penguin Books 2007
034

Fooled by Randomness is a work of non-fiction, but certain names of non-public
figures have been changed, and some of the private individuals described
are fictionalized or composite portraits.

Printed in England by Clays Ltd, St Ives plc

ISBN: 978–0–141–03148–4

www.greenpenguin.co.uk

MIX
Paper from
responsible sources
FSC
www.fsc.org FSC™ C018179

Penguin Books is committed to a sustainable
future for our business, our readers and our planet.
This book is made from Forest Stewardship
Council™ certified paper.

To my mother,
Minerva Ghosn Taleb

PREFACE

TAKING KNOWLEDGE
LESS SERIOUSLY

This book is the synthesis of, on one hand, the no-nonsense practitioner of uncertainty who spent his professional life trying to resist being fooled by randomness and trick the emotions associated with probabilistic outcomes and, on the other, the aesthetically obsessed, literature-loving human being willing to be fooled by any form of nonsense that is polished, refined, original, and tasteful. I am not capable of avoiding being the fool of randomness; what I can do is confine it to where it brings some aesthetic gratification.

This comes straight from the gut; it is a personal essay primarily discussing its author's thoughts, struggles, and observations connected to the practice of risk taking, not exactly a treatise, and certainly, god forbid, not a piece of scientific reporting. It was written for fun and it aims to be read (principally) for, and with, pleasure. Much has been written about our biases (acquired or genetic) in dealing with randomness over the past decade. The rules while writing the first edition of this book had been to avoid discussing (a) anything that I did not either personally witness on the topic or develop independently, and (b) anything that I have not distilled well enough to be able to write on the subject with only the slightest effort. Everything that remotely felt like work was out. I had to purge from the text passages that seemed to come from a

visit to the library, including the scientific name dropping. I tried
to use no quote that did not naturally spring from my memory and
did not come from a writer whom I had intimately frequented
over the years (I detest the practice of random use of borrowed
wisdom—much on that later). *Aut tace aut loquere meliora silencio*
(only when the words outperform silence).

These rules remain intact. But sometimes life requires compro-
mises: Under pressure from friends and readers I have added to
the present edition a series of nonintrusive endnotes referring to
the related literature. I have also added new material to most
chapters, most notably in Chapter 11, which altogether has resulted in an expansion of the book by more than a third.

Adding to the Winner

I hope to make this book organic—by, to use traders' lingo, "adding
to the winner"—and let it reflect my personal evolution instead of
holding on to these new ideas and putting them into a new book
altogether. Strangely, I gave considerably more thought to some
sections of this book *after* the publication than I had before, particularly in two separate areas: (a) the mechanisms by which our
brain sees the world as less, far less, random that it actually is, and
(b) the "fat tails," that wild brand of uncertainty that causes large
deviations (rare events explain more and more of the world we
live in, but at the same time remain as counterintuitive to us as
they were to our ancestors). The second version of this book re-
flects this author's drift into becoming a little less of a student of
uncertainty (we can learn so little about randomness) and more of
a researcher into how people are fooled by it.

Another phenomenon: the transformation of the author by his
own book. As I increasingly started living this book *after* the initial
composition, I found luck in the most unexpected of places. It is
as if there were two planets: the one in which we actually live and
the one, considerably more deterministic, on which people are

convinced we live. It is as simple as that: Past events will *always* look less random than they were (it is called the *hindsight bias*). I would listen to someone's discussion of his own past realizing that much of what he was saying was just backfit explanations concocted ex post by his deluded mind. This became at times unbearable: I could feel myself looking at people in the social sciences (particularly conventional economics) and the investment world as if they were deranged subjects. Living in the real world may be painful particularly if one finds statements more informative about the people making them than the intended message: I picked up *Newsweek* this morning at the dentist's office and read a journalist's discussion of a prominent business figure, particularly his ability in "timing moves" and realized how I was making a list of the biases in the journalist's mind rather than getting the intended information in the article itself, which I could not possibly take seriously. (Why don't most journalists end up figuring out that they know much less than they think they know? Scientists investigated half a century ago the phenomena of "experts" not learning about their past failings. You can mispredict everything for all your life yet think that you will get it right next time.)

Insecurity and Probability

I believe that the principal asset I need to protect and cultivate is my deep-seated intellectual insecurity. My motto is *"my principal activity is to tease those who take themselves and the quality of their knowledge too seriously."* Cultivating such insecurity in place of intellectual confidence may be a strange aim—and one that is not easy to implement. To do so we need to purge our minds of the recent tradition of intellectual certainties. A reader turned pen pal made me rediscover the sixteenth-century French essayist and professional introspector Montaigne. I got sucked into the implications of the difference between Montaigne and Descartes—and how we strayed by following the latter's quest for certitudes. We surely closed our minds by following Descartes' model of formal thinking rather than Montaigne's brand of vague and informal (but critical) judgment. Half a millennium later the severely introspecting and insecure Montaigne stands tall as a role model for the modern thinker. In addition, the man had exceptional courage: It certainly takes bravery to remain skeptical; it takes inordinate courage to introspect, to confront oneself, to accept one's limitations—scientists are seeing more and more evidence that we are specifically designed by mother nature to fool ourselves.

There are many intellectual approaches to probability and risk—"probability" means slightly different things to people in different disciplines. In this book it is tenaciously qualitative and literary as opposed to quantitative and "scientific" (which explains the warnings against economists and finance professors as they tend to firmly believe that they know something, and something useful at that). It is presented as flowing from Hume's Problem of Induction (or Aristotle's inference to the general) as opposed to the paradigm of the gambling literature. In this book probability is principally a branch of applied skepticism, not an engineering discipline (in spite of all the self-important mathematical treatment of the subject matter, problems related to the calculus of probability rarely merit to transcend the footnote).

How? Probability is not a mere computation of odds on the dice or more complicated variants; it is the acceptance of the lack of certainty in our knowledge and *the development of methods for dealing with our ignorance.* Outside of textbooks and casinos, probability almost *never* presents itself as a mathematical problem or a brain teaser. Mother nature does not tell you how many holes there are on the roulette table, nor does she deliver problems in a textbook way (in the real world one has to guess the problem more than the solution). In this book, considering that alternative outcomes could have taken place, that the world could have been different, is the core of probabilistic thinking. As a matter of fact,

I spent all my career attacking the *quantitative* use of probability. While Chapters 13 and 14 (dealing with skepticism and stoicism) are to me the central ideas of the book, most people focused on the examples of miscomputation of probability in Chapter 11 (clearly and by far the least original chapter of the book, one in which I compressed all the literature on probability biases). In addition, while we may have some understanding of the probabilities in the hard sciences, particularly in physics, we don't have much of a clue in the social "sciences" like economics, in spite of the fanfares of experts.

Vindicating (Some) Readers

I have tried to make the minimum out of my occupation of mathematical trader. The fact that I operate in the markets serves only as an inspiration—it does not make this book (as many thought it was) a guide to market randomness any more than the *Iliad* should be interpreted as a military instruction manual. Only three out of fourteen chapters have a financial setting. Markets are a mere special case of randomness traps—but they are by far the most interesting as luck plays a very large role in them (this book would have been considerably shorter if I were a taxidermist or a translator of chocolate labels). Furthermore, the kind of luck in finance is of the kind that nobody understands but most operators *think* they understand, which provides us a magnification of the biases. I have tried to use my market analogies in an illustrative way as I would in a dinner conversation with, say, a cardiologist with intellectual curiosity (I used as a model my second-generation friend Jacques Merab).

I received large quantities of electronic mail on the first version of the book, which can be an essayist's dream as such dialectic provides ideal conditions for the rewriting of the second version. I expressed my gratitude by answering (once) each one of them. Some of the answers have been inserted back into the text in the different chapters. Being often seen as an iconoclast I was looking forward to getting the angry letters of the type "who are you to judge Warren Buffett" or "you are envious of his success"; instead it was disappointing to see most of the trashing going anonymously to amazon.com (there is no such thing as bad publicity: Some people manage to promote your work by insulting it).

The consolation for the lack of attacks was in the form of letters from people who felt vindicated by the book. The most rewarding letters were the ones from people who did not fare well in life, through no fault of their own, who used the book as an argument with their spouse to explain that they were less lucky (not less skilled) than their brother-in-law. The most touching letter came from a man in Virginia who within a period of a few months lost his job, his wife, his fortune, was put under investigation by the redoubtable Securities and Exchange Commission, and progressively felt good for acting stoically. A correspondence with a reader who was hit with a black swan, the unexpected large-impact random event (the loss of a baby) caused me to spend some time dipping into the literature on adaptation after a severe random event (not coincidentally also dominated by Daniel Kahneman, the pioneer of the ideas on irrational behavior under uncertainty). I have to confess that I never felt really particularly directly of service to anyone being a trader (except myself); it felt elevating and *useful* being an essayist.

All or None

A few confusions with the message in this book. Just as our brain does not easily make out probabilistic shades (it goes for the oversimplifying "all-or-none"), it was hard to explain that the idea here was that "it is more random than we think" rather than "it is all random." I had to face the "Taleb, as a skeptic, thinks everything is random and successful people are just lucky." The Fooled by Randomness symptom even affected a well-publicized Cambridge

Union Debate as my argument "*Most* City Hotshots are Lucky Fools" became "*All* City Hotshots are Lucky Fools" (clearly I lost the debate to the formidable Desmond Fitzgerald in one of the most entertaining discussions in my life—I was even tempted to switch sides!). The same delusion of mistaking irreverence for arrogance (as I noticed with my message) makes people confuse skepticism for nihilism.

Let me make it clear here: Of course chance favors the prepared! Hard work, showing up on time, wearing a clean (preferably white) shirt, using deodorant, and some such conventional things contribute to success—they are certainly necessary but may be insufficient as they do not *cause* success. The same applies to the conventional values of persistence, doggedness and perseverance: *necessary, very necessary.* One needs to go out and buy a lottery ticket in order to win. Does it mean that the work involved in the trip to the store *caused* the winning? Of course skills count, but they do count less in highly random environments than they do in dentistry.

No, I am not saying that what your grandmother told you about the value of work ethics is wrong! Furthermore, as most successes are caused by very few "windows of opportunity," failing to grab one can be deadly for one's career. Take your luck!

Notice how our brain sometimes gets the arrow of causality backward. Assume that good qualities *cause* success; based on that assumption, even though it seems intuitively correct to think so, the fact that every intelligent, hardworking, persevering person becomes successful does not imply that every successful person is necessarily an intelligent, hardworking, persevering person (it is remarkable how such a primitive logical fallacy—*affirming the consequent*—can be made by otherwise very intelligent people, a point I discuss in this edition as the "two systems of reasoning" problem).

There is a twist in research on success that has found its way into the bookstores under the banner of advice on: "these are the millionaires' traits that you need to have if you want to be just like those successful people." One of the authors of the misguided *The Millionaire Next Door* (that I discuss in Chapter 8) wrote another even more foolish book called *The Millionaire Mind*. He observes that in the representative cohort of more than a thousand millionaires whom he studied most did not exhibit high intelligence in their childhood and infers that it is not your endowment that makes you rich—but rather hard work. From this, one can naively infer that chance plays no part in success. My intuition is that if millionaires are close in attributes to the average population, then I would make the more disturbing interpretation that it is because luck played a part. Luck is democratic and hits everyone regardless of original skills. The author notices variations from the general population in a few traits like tenacity and hard work: another confusion of the *necessary* and the causal. That all millionaires were persistent, hardworking people does not make persistent hard workers become millionaires: Plenty of unsuccessful entrepreneurs were persistent, hardworking people. In a textbook case of naive empiricism, the author also looked for traits these millionaires had in common and figured out that they shared a taste for risk taking. Clearly risk taking is necessary for large success— but it is also necessary for failure. Had the author done the same study on bankrupt citizens he would certainly have found a predilection for risk taking.

I was asked to "back up the claims" in the book with the "supply of data," graphs, charts, diagrams, plots, tables, numbers, recommendations, time series, etc., by some readers (and by *me-too* publishers before I was lucky to find Texere). This text is a series of logical thought experiments, not an economics term paper; logic does not require empirical verification (again there is what I call a "round-trip fallacy": It is a mistake to use, as journalists and some economists do, statistics without logic, but the reverse does not hold: It is not a mistake to use logic without statistics). If I write that I doubt that my neighbor's success is devoid of some measure, small or large, of luck, owing to the randomness in his profession, I do not need to "test" it—the Russian roulette thought experiment suffices. All I need is to show that there exists an alternative explanation to the theory that he is a genius. My approach is to manufacture a cohort of intellectually challenged persons and show how a small minority can evolve into successful businessmen—but these are the ones who will be visible. I am not saying that Warren Buffett is not skilled; only that a large population of random investors will *almost necessarily* produce someone with his track records *just by luck*.

Missing a Hoax

I was also surprised at the fact that in spite of the book's aggressive warning against media journalism I was invited to television and radio shows in both North America and Europe (including a hilarious *dialogue de sourds* on a Las Vegas radio station where the interviewer and I were running two parallel conversations). Nobody protected me from myself and I accepted the interviews. Strangely, one needs to use the press to communicate the message that the press is toxic. I felt like a fraud coming up with vapid sound bites, but had fun at it.

It may be that I was invited because the mainstream media interviewers did not read my book or understand the insults (they don't "have the time" to read books) and the nonprofit ones read it too well and felt vindicated by it. I have a few anecdotes: A famous television show was told that "this guy Taleb believes that stock analysts are just random forecasters" so they seemed eager to have me present my ideas on the program. However, their condition was that I make three stock recommendations to prove my "expertise." I didn't attend and missed the opportunity for a great hoax by discussing three stocks selected randomly and fitting a well-sounding explanation to my selection.

On another television show I mentioned that "people think that there is a story when there is none" as I was discussing the random character of the stock market and the backfit logic one always sees in events after the fact. The anchor immediately interjected: "There was a story about Cisco this morning. Can you comment on that?" The best: When invited to an hour-long discussion on a financial radio show (they had not read Chapter 11), I was told a few minutes before to refrain from discussing the ideas in this book because I was invited to talk about trading and not about randomness (another hoax opportunity certainly, but I was too unprepared for it and walked out before the show started).

Most journalists do not take things too seriously: After all, this business of journalism is about pure entertainment, not a search for truth, particularly when it comes to radio and television. The trick is to stay away from those who do not seem to know that they are just entertainers (like George Will, who will appear in Chapter 2) and actually believe that they are *thinkers*.

Another problem was in the interpretation of the message in the media: This guy Nassim thinks that markets are random, *hence they are going lower*, which made me the unwilling bearer of catastrophic messages. Black swans, those rare and unexpected deviations, can be both good and bad events.

However, media journalism is less standardized than it appears; it attracts a significant segment of thoughtful people who manage to extricate themselves from the commercial sound bite–driven system and truly care about the message rather than just catching the public's attention. One naive observation from my conversations with Kojo Anandi (NPR), Robin Lustig (BBC), Robert Scully (PBS), and Brian Lehrer (WNYC) is that the nonprofit journalist is altogether another intellectual breed. Casually, the quality of the discussion correlates inversely with the luxury of the studios: WNYC, where I felt that Brian Lehrer was making the greatest effort at getting into the arguments, operates out of the shabbiest offices I have seen this side of Kazakhstan.

One final comment on the style. I elected to keep the style of this book as idiosyncratic as it was in the first edition. *Homo sum*, good and bad. I am fallible and see no reason to hide my minor flaws if they are part of my personality no more than I feel the need to wear a wig when I have my picture taken or borrow someone else's nose when I show my face. Almost all the book editors who read the draft recommended changes at the sentence level (to make my style "better") and in the structure of the text (in the organization of chapters); I ignored almost all of them and found out that none of the readers thought them necessary—as a matter of fact, I find that injecting the personality of the author (imperfections included) enlivens the text. Does the book industry suffer from the classical "expert problem" with the buildup of rules of thumb that do not have empirical validity? More than a hundred thousand readers later I am discovering that books are not written for book editors.

ACKNOWLEDGMENTS
FOR THE UPDATED SECOND EDITION

Out of the Library

The book helped me break out of my intellectual isolation (not being a full-time academic offers plenty of benefits, such as independence and the avoidance of the dull parts of the process, but it comes at the cost of seclusion). I made many interesting dinner companions and pen pals among lucid thinkers through the first edition, and, thanks to them, I was able to make a second pass on some of the topics. In addition, I have gotten closer to my dream life thanks to the stimulation of discussion with people who share my interests; I feel that I need to pay the book back for that. There seems to be some evidence that conversations and correspondence with intelligent people is a better engine for personal edification than plain library-ratting (human warmth: Something in our nature may help us grow ideas while dealing and socializing with other people). Somehow there was the *pre-* and *post-Fooled* life. While the acknowledgments for the first edition hold more than ever, I would like to add here my newly incurred debt.

Shrinking the World

I first met Robert Shiller in person as we were seated next to each other at a breakfast panel discussion. I found myself inadvertently eating all the fruits on his plate and drinking his coffee and water, leaving him with the muffins and other unfashionable food (and nothing to drink). He did not complain (he may have not noticed). I did not know Shiller when I featured him in the first edition and was surprised by his accessibility, his humility, and his charm (by some heuristic one does not expect people who have vision to be also personable). He later drove me to a bookstore in New Haven, showed me *Flatland*, a scientific parable dealing with physics that he read when he was in high school, and told me to keep this book as it was in the first edition: short, personal, as close to a novel as possible, something I kept in mind throughout the exercise of this reworking (he tried to convince me to not do this second edition, I begged him to do a second one of his own *Irrational Exuberance*, be it only for my own consumption; I think that I won both points). Books have bubble dynamics of the type discussed in Chapter 10, a matter that makes an extra edition of an existing book far more likely to break through the critical point than a new one (network externalities make religions and fads fare incrementally better in their second editions than brand-new ones). The physicist and crash theorist Didier Sornette provided me with convincing arguments for the effectiveness of a second version; we are surprised that book publishers who thrive on informational cascades are not conscious of the point.

During much of the rewriting of this book I was under the energizing influence of two intense dinner conversations in Italy with Daniel Kahneman, which had the effect of "pushing" me to the next critical point of intellectual drive, after I saw that his work went so much deeper than mere rational choice under uncertainty. I am certain that his influence on economics (including the Nobel medal) focused people away from the breadth and depth and the general applicability of his discoveries. Economics is boring stuff, but *His work matters* I kept telling myself, not just because he is an empiricist, not just because of the contrast of the relevance of his work (and personality) with those of the other recent Nobel economists, but because of its far-reaching implications on far worthier questions: (a) He and Amos Tversky helped stand on its head the notion of man that we owe to the dogmatic rationalism of the Hellenistic age and which held for twenty-three centuries, with all the damaging consequences that we know of now; (b) Kahneman's important work is on utility theory (in its different stages) with consequences on such significant things as happiness. Now understanding happiness is a *real* pursuit.

I had lengthy discussions with Terry Burnham, the biologist and evolutionary economist and co-author of *Mean Genes*, that unpretentious introduction to evolutionary psychology, who coincidentally turned out to be best friends with Jamil Baz, the childhood friend who was my sounding board with my early introspections on randomness two decades ago. Peter McBurney got me involved with the Artificial Intelligence community, which seems to fuse together the fields of philosophy, cognitive neuroscience, mathematics, economics, and logic. He and I started a voluminous correspondence on the various theories of rationality. Michael Schrage, one of my reviewers, is the epitome of the modern (hence scientific) intellectual—he has a knack of reading everything that seems to matter. He offers the conversation of a true intellectual, shielded from the straitjacket of academic pressures. Ramaswami Ambarish and Lester Siegel showed me (with their suspiciously unnoticed work) that if we are fooled by randomness with respect to plain performance, then performance differential is even harder to pin down. The writer Malcolm Gladwell sent me into some interesting parts of the literature on intuition and self-knowledge. Art De Vany, the insightful and brilliantly colorful economist who specializes in nonlinearities and rare events, started his introductory letter to me with the shibboleth "I despise textbooks." It is encouraging to see someone with such depth in his thinking who can also have fun in life. The economist William Easterly showed me that randomness contributed to illusionary causes in economic development. He liked the link between being a skeptical empiricist and disliking monopolies on knowledge by institutions like governments and universities. I am grateful to Hollywood agent Jeff Berg, an enthusiastic reader, for his insights on the wild type of uncertainty that prevails in the media business. I have to thank the book for allowing me to have insightful dinner discussions with Jack Schwager, who seems to have thought of some of the problems longer than anybody alive.

Thank You, Google

The following people have provided me with help on this text. I was very fortunate to have Andreea Munteanu as an incisive reader and valuable sounding board; she spent hours away from her impressive derivatives job checking the integrity of the references on Google. Amanda Gharghour also helped with the search. I was also lucky to have Gianluca Monaco as the Italian translator; he found mistakes in the text that it would have taken me a century to detect (a cognitive scientist and book-translator-turned-student-of-mathematical-finance, he called up the publisher and appointed himself the translator). My collaborator, the philosopher of science Avital Pilpel, provided me with invaluable help with technical probability discussions. Elie Ayache, another Levantine-trader-mathematician-physicist-turned-philosopher-of-science-probability-markets (though without the neurobiology), made me spend numerous hours at Borders Books in both the philosophy section and the science section. Flavia Cymbalista, Sole Marittimi (now Riley), Paul Wilmott, Mark Spitznagel, Gur Huberman, Tony Glickman, Winn Martin, Alexander Reisz, Ted Zink, Andrei Pokrovsky, Shep Davis, Guy Riviere, Eric Schoenberg, and Marco Di Martino provided comments on the text. George Martin was, as usual, an invaluable sounding board. The readers Carine Chichereau, Bruce Bellner, and Illias Katsounis, gracefully e-mailed me extensive errata. I thank Cindy,

Sarah, and Alexander for support and the reminder that there are other things than probability and uncertainty.

I also have to thank my second home, the Courant Institute of Mathematical Sciences, for providing me with the right atmosphere to pursue my interests and teach and coach students while retaining my intellectual independence, particularly Jim Gatheral, who took the habit of heckling me while co-teaching the class. I am indebted to Paloma's Donald Sussman and Tom Witz for their unusual insights; I am truly impressed by their heroic ability to understand the "black swan." I also thank the Empirica members (we ban the use of the word *employees*) for fostering a climate of fierce and ruthless, truly cut-throat intellectual debate in the office. My coworkers make sure that not a single comment on my part can go without some sort of challenge.

I insist once again that without David Wilson and Myles Thompson this book would have never been initially published. But without Will Murphy, Daniel Menaker, and Ed Klagsbrun, who revived this book, it would have been dead. I thank Janet Wygal for her thoroughness (and patience) and Fleetwood Robbins for his assistance. Given their zeal, I doubt that many mistakes are left; however, those that remain are mine.

CONTENTS

Preface vii

Acknowledgments for the Updated Second Edition xix

Chapter Summaries xxxv

Prologue xxxix

PART I: SOLON'S WARNING

•

Skewness, Asymmetry, Induction

One   **IF YOU'RE SO RICH, WHY AREN'T YOU SO SMART?** 5

NERO TULIP 5

Hit by Lightning 5

Temporary Sanity 6

Modus Operandi 9

No Work Ethics 11

There Are Always Secrets 12

JOHN THE HIGH-YIELD TRADER 12

An Overpaid Hick 14

THE RED-HOT SUMMER   17
   Serotonin and Randomness   18
YOUR DENTIST IS RICH, VERY RICH 20

Two   **A BIZARRE ACCOUNTING METHOD**   22
ALTERNATIVE HISTORY   22
   Russian Roulette   23
   Possible Worlds   24
   An Even More Vicious Roulette   26
SMOOTH PEER RELATIONS   27
   Salvation via Aeroflot   29
   Solon Visits Regine's Nightclub   30
GEORGE WILL IS NO SOLON:
ON COUNTERINTUITIVE TRUTHS   33
   Humiliated in Debates   36
   A Different Kind of Earthquake   37
   Proverbs Galore   39
   Risk Managers   40
   Epiphenomena   41

Three   **A MATHEMATICAL MEDITATION
ON HISTORY**   43
   Europlayboy Mathematics   43
   The Tools   44
   Monte Carlo Mathematics   47
FUN IN MY ATTIC   49
   Making History   49
   Zorglubs Crowding the Attic   50
   Denigration of History   51
   The Stove Is Hot   52
   Skills in Predicting Past History   55

My Solon 56

DISTILLED THINKING ON YOUR
PALMPILOT 58

   Breaking News 58

   Shiller Redux 61

   Gerontocracy 63

PHILOSTRATUS IN MONTE CARLO:
ON THE DIFFERENCE BETWEEN NOISE
AND INFORMATION 64

Four   **RANDOMNESS, NONSENSE, AND
THE SCIENTIFIC INTELLECTUAL** 70

RANDOMNESS AND THE VERB 70

   Reverse Turing Test 72

   The Father of All Pseudothinkers 74

MONTE CARLO POETRY 75

Five   **SURVIVAL OF THE LEAST FIT—CAN EVOLUTION
BE FOOLED BY RANDOMNESS?** 79

CARLOS THE EMERGING-MARKETS WIZARD 79

   The Good Years 82

   Averaging Down 83

   Lines in the Sand 84

JOHN THE HIGH-YIELD TRADER 86

   *The Quant Who Knew Computers
and Equations* 87

   The Traits They Shared 90

A REVIEW OF MARKET FOOLS OF RANDOMNESS
CONSTANTS 91

NAIVE EVOLUTIONARY THEORIES 94

   Can Evolution Be Fooled by Randomness? 96

Six   **SKEWNESS AND ASYMMETRY**   97

THE MEDIAN IS NOT THE MESSAGE   97

BULL AND BEAR ZOOLOGY   99

An Arrogant Twenty-nine-year-old Son   102

Rare Events   103

Symmetry and Science   105

ALMOST EVERYBODY IS ABOVE AVERAGE   106

THE RARE-EVENT FALLACY   108

The Mother of All Deceptions   108

Why Don't Statisticians Detect Rare Events?   112

A Mischievous Child Replaces
the Black Balls   113

Seven   **THE PROBLEM OF INDUCTION**   116

FROM BACON TO HUME   116

Cygnus Atratus   117

Niederhoffer   117

SIR KARL'S PROMOTING AGENT   121

Location, Location   125

Popper's Answer   126

Open Society   128

Nobody Is Perfect   129

Induction and Memory   130

Pascal's Wager   130

THANK YOU, SOLON   131

PART II MONKEYS ON TYPEWRITERS

●

Survivorship and Other Biases

IT DEPENDS ON THE NUMBER OF MONKEYS   136

VICIOUS REAL LIFE   137

THIS SECTION   137

Eight   **TOO MANY MILLIONAIRES NEXT DOOR**   139

HOW TO STOP THE STING OF FAILURE   139

Somewhat Happy   139

Too Much Work   140

You're a Failure   141

DOUBLE SURVIVORSHIP BIASES   143

More Experts   143

Visibility Winners   145

It's a Bull Market   145

A GURU'S OPINION   147

Nine   **IT IS EASIER TO BUY AND SELL**

THAN FRY AN EGG   149

FOOLED BY NUMBERS   151

Placebo Investors   151

Nobody Has to Be Competent   153

Regression to the Mean   155

Ergodicity   156

LIFE IS COINCIDENTAL   157

The Mysterious Letter   157

An Interrupted Tennis Game   158

Reverse Survivors   159

The Birthday Paradox 159

It's a Small World! 159

Data Mining, Statistics, and Charlatanism 160

The Best Book I Have Ever Read! 161

The Backtester 162

A More Unsettling Extension 164

The Earnings Season: Fooled by the Results 164

COMPARATIVE LUCK 165

Cancer Cures 166

Professor Pearson Goes to Monte Carlo (Literally):
Randomness Does Not Look Random! 168

The Dog That Did Not Bark: On Biases
in Scientific Knowledge 170

I HAVE NO CONCLUSION 170

Ten   **LOSER TAKES ALL—ON THE NONLINEARITIES**
OF LIFE 172

THE SANDPILE EFFECT 172

Enter Randomness 174

Learning to Type 175

MATHEMATICS INSIDE AND OUTSIDE THE REAL
WORLD 176

The Science of Networks 178

Our Brain 179

Buridan's Donkey or the Good Side
of Randomness 179

WHEN IT RAINS, IT POURS 180

Eleven   **RANDOMNESS AND OUR MIND: WE ARE**
PROBABILITY BLIND 182

PARIS OR THE BAHAMAS? 182

SOME ARCHITECTURAL
CONSIDERATIONS 183

BEWARE THE PHILOSOPHER BUREAUCRAT 185

Satisficing 186

FLAWED, NOT JUST IMPERFECT 187

Kahneman and Tversky 187

WHERE IS NAPOLEON WHEN WE
NEED HIM? 190

*"I'm As Good As My Last Trade"
and Other Heuristics* 191

Degree in a Fortune Cookie 194

Two Systems of Reasoning 196

WHY WE DON'T MARRY THE FIRST DATE 197

Our Natural Habitat 198

Fast and Frugal 200

Neurobiologists Too 201

Kafka in a Courtroom 203

An Absurd World 205

*Examples of Biases in Understanding
Probability* 206

We Are Option Blind 207

PROBABILITIES AND THE MEDIA
(MORE JOURNALISTS) 210

CNBC at Lunchtime 211

You Should Be Dead by Now 212

The Bloomberg Explanations 213

Filtering Methods 216

We Do Not Understand Confidence Levels 216

An Admission 218

PART III   WAX IN MY EARS

•

Living with Randomitis

I AM NOT SO INTELLIGENT   222

WITTGENSTEIN'S RULER   223

THE ODYSSEAN MUTE COMMAND   224

Twelve   **GAMBLERS' TICKS AND PIGEONS IN A BOX**   226

TAXI-CAB ENGLISH AND CAUSALITY   226

THE SKINNER PIGEON EXPERIMENT   229

PHILOSTRATUS REDUX   231

Thirteen   **CARNEADES COMES TO ROME: ON PROBABILITY AND SKEPTICISM**   234

CARNEADES COMES TO ROME   235

Probability, the Child of Skepticism   237

MONSIEUR DE NORPOIS' OPINIONS   238

Path Dependence of Beliefs   239

COMPUTING INSTEAD OF THINKING   241

FROM FUNERAL TO FUNERAL   244

Fourteen   **BACCHUS ABANDONS ANTONY**   245

NOTES ON JACKIE O.'S FUNERAL   247

RANDOMNESS AND PERSONAL ELEGANCE   249

Epilogue   **SOLON TOLD YOU SO**   250

Beware the London Traffic Jams   250

Postscript **THREE AFTERTHOUGHTS IN THE
SHOWER** 253

FIRST THOUGHT: THE INVERSE SKILLS
PROBLEM 254

SECOND THOUGHT: ON SOME ADDITIONAL
BENEFITS OF RANDOMNESS 257

Uncertainty and Happiness 257

The Scrambling of Messages 261

THIRD THOUGHT: STANDING ON ONE LEG 262

Acknowledgments for the First Edition 263

A Trip to the Library: Notes and Reading Recommendations 267

Notes 269

References 293

Index 307

CHAPTER SUMMARIES

ONE: IF YOU'RE SO RICH, WHY AREN'T YOU SO SMART?
An illustration of the effect of randomness on social pecking order and jealousy, through two characters of opposite attitudes. On the concealed rare event. How things in modern life may change rather rapidly, except, perhaps, in dentistry.

TWO: A BIZARRE ACCOUNTING METHOD
On alternative histories, a probabilistic view of the world, intellectual fraud, and the randomness wisdom of a Frenchman with steady bathing habits. How journalists are bred to not understand random series of events. Beware borrowed wisdom: How almost all great ideas concerning random outcomes are against conventional sapience. On the difference between correctness and intelligibility.

THREE: A MATHEMATICAL MEDITATION ON HISTORY
On Monte Carlo simulation as a metaphor for understanding a sequence of random historical events. On randomness and artificial history. Age is beauty, almost always, and the new and the young are generally toxic. Send your history professor to an introductory class on sampling theory.

FOUR: RANDOMNESS, NONSENSE, AND THE SCIENTIFIC INTELLECTUAL

On extending the Monte Carlo generator to produce artificial thinking and compare it with rigorous nonrandom constructs. The science wars enter the business world. Why the aesthete in me loves to be fooled by randomness.

FIVE: SURVIVAL OF THE LEAST FIT—CAN EVOLUTION BE FOOLED BY RANDOMNESS?

A case study on two rare events. On rare events and evolution. How "Darwinism" and evolution are concepts that are misunderstood in the nonbiological world. Life is not continuous. How evolution will be fooled by randomness. A prolegomenon for the problem of induction.

SIX: SKEWNESS AND ASYMMETRY

We introduce the concept of skewness: Why the terms "bull" and "bear" have limited meaning outside of zoology. A vicious child wrecks the structure of randomness. An introduction to the problem of epistemic opacity. The penultimate step before the problem of induction.

SEVEN: THE PROBLEM OF INDUCTION

On the chromodynamics of swans. Taking Solon's warning into some philosophical territory. How Victor Niederhoffer taught me empiricism; I added deduction. Why it is not scientific to take science seriously. Soros promotes Popper. That bookstore on Eighteenth Street and Fifth Avenue. Pascal's wager.

EIGHT: TOO MANY MILLIONAIRES NEXT DOOR

Three illustrations of the survivorship bias. Why very few people should live on Park Avenue. The millionaire next door has very flimsy clothes. An overcrowding of experts.

NINE: IT IS EASIER TO BUY AND SELL THAN FRY AN EGG

Some technical extensions of the survivorship bias. On the distribution of "coincidences" in life. It is preferable to be lucky than competent (but you can be caught). The birthday paradox. More charlatans (and more journalists). How the researcher with work ethics can find just about anything in data. On dogs not barking.

TEN: LOSER TAKES ALL—ON THE NONLINEARITIES OF LIFE

The nonlinear viciousness of life. Moving to Bel Air and acquiring the vices of the rich and famous. Why Microsoft's Bill Gates may not be the best in his business (but please do not inform him of such a fact). Depriving donkeys of food.

ELEVEN: RANDOMNESS AND OUR MIND: WE ARE PROBABILITY BLIND

On the difficulty of thinking of your vacation as a linear combination of Paris and the Bahamas. Nero Tulip may never ski in the Alps again. Do not ask bureaucrats too many questions. A Brain Made in Brooklyn. We need Napoleon. Scientists bowing to the King of Sweden. A little more on journalistic pollution. Why you may be dead by now.

TWELVE: GAMBLERS' TICKS AND PIGEONS IN A BOX

On gamblers' ticks crowding up my life. Why bad taxi-cab English can help you make money. How I am the fool of all fools, except that I am aware of it. Dealing with my genetic unfitness. No boxes of chocolate under my trading desk.

THIRTEEN: CARNEADES COMES TO ROME: ON PROBABILITY AND SKEPTICISM

Cato the censor sends Carneades packing. Monsieur de Norpois does not remember his old opinions. Beware the scientist. Marrying ideas. The same Robert Merton putting the author on the map. Science evolves from funeral to funeral.

FOURTEEN: BACCHUS ABANDONS ANTONY

Montherlant's death. Stoicism is not the stiff upper lip, but the illusion of victory of man against randomness. It is so easy to be heroic. Randomness and personal elegance.

PROLOGUE

MOSQUES IN THE CLOUDS

This book is about luck disguised and perceived as nonluck (that is, skills) and, more generally, randomness disguised and perceived as non-randomness (that is, determinism). It manifests itself in the shape of the *lucky fool*, defined as a person who benefited from a disproportionate share of luck but attributes his success to some other, generally very precise, reason. Such confusion crops up in the most unexpected areas, even science, though not in such an accentuated and obvious manner as it does in the world of business. It is endemic in politics, as it can be encountered in the shape of a country's president discoursing on the jobs that "he" created, "his" recovery, and "his predecessor's" inflation.

We are still very close to our ancestors who roamed the savannah. The formation of our beliefs is fraught with superstitions— even today (I might say, especially today). Just as one day some primitive tribesman scratched his nose, saw rain falling, and developed an elaborate method of scratching his nose to bring on the much-needed rain, we link economic prosperity to some rate cut by the Federal Reserve Board, or the success of a company with the appointment of the new president "at the helm." Bookstores are full of biographies of successful men and women presenting their specific explanation on how they made it big in life (we have an expression, "the right time and the right place," to weaken whatever conclusion can be inferred from them). This confusion strikes people of different persuasions; the literature professor invests a deep meaning into a mere coincidental occurrence of word patterns, while the economist proudly detects "regularities" and "anomalies" in data that are plain random.

At the cost of appearing biased, I have to say that the literary mind can be intentionally prone to the confusion between *noise* and *meaning*, that is, between a randomly constructed arrangement and a precisely intended message. However, this causes little harm; few claim that art is a tool of investigation of the Truth— rather than an attempt to escape it or make it more palatable. Symbolism is the child of our inability and unwillingness to accept randomness; we give meaning to all manner of shapes; we detect human figures in inkblots. *I saw mosques in the clouds* announced Arthur Rimbaud, the nineteenth-century French symbolic poet. This interpretation took him to "poetic" Abyssinia (in East Africa), where he was brutalized by a Christian Lebanese slave dealer, contracted syphilis, and lost a leg to gangrene. He gave up poetry in disgust at the age of nineteen, and died anonymously in a Marseilles hospital ward while still in his thirties. But it was too late.

European intellectual life developed what seems to be an irreversible taste for symbolism—we are still paying its price, with psychoanalysis and other fads.

Regrettably, some people play the game too seriously; they are paid to read too much into things. All my life I have suffered the conflict between my love of literature and poetry and my profound allergy to most teachers of literature and "critics." The French thinker and poet Paul Valery was surprised to listen to a commentary of his poems that found meanings that had until then escaped him (of course, it was pointed out to him that these were intended by his subconscious).

More generally, we underestimate the share of randomness in about everything, a point that may not merit a book—except when it is the specialist who is the fool of all fools. Disturbingly, science has only recently been able to handle randomness (the growth in available information has been exceeded only by the expansion of noise). Probability theory is a young arrival in mathematics; probability applied to practice is almost nonexistent as a discipline. In addition we seem to have evidence that what is called "courage" comes from an underestimation of the share of randomness in things rather than the more noble ability to stick one's neck out for a given belief. In my experience (and in the scientific literature), economic "risk takers" are rather the victims of delusions (leading to overoptimism and overconfidence with their underestimation of possible adverse outcomes) than the opposite. Their "risk taking" is frequently randomness foolishness.

Consider the left and the right columns of Table P.1 (page xliii). The best way to summarize the major thesis of this book is that it addresses situations (many of them tragicomical) where the left column is mistaken for the right one. The subsections also illustrate the key areas of discussion on which this book will be based.

The reader may wonder whether the opposite case might not deserve some attention, that is, the situations where nonrandomness is mistaken for randomness. Shouldn't we be concerned with situations where patterns and messages may have been ignored? I have two answers. First, I am not overly worried about the existence of undetected patterns. We have been reading lengthy and complex messages in just about any manifestation of nature that presents jaggedness (such as the palm of a hand, the residues at the bottom of Turkish coffee cups, etc.). Armed with home supercomputers and chained processors, and helped by complexity and "chaos" theories, the scientists, semiscientists, and pseudoscientists will be able to find portents. Second, we need to take into account the costs of mistakes; in my opinion, mistaking the right column for the left one is not as costly as an error in the opposite direction. Even popular opinion warns that bad information is worse than no information at all.

However interesting these areas could be, their discussion would be a tall order. There is one world in which I believe the habit of mistaking luck for skill is most prevalent—and most conspicuous—and that is the world of markets. By luck or misfortune, that is the world in which I have operated most of my adult life. It is what I know best. In addition, economic life presents the best (and most entertaining) laboratory for the understanding of these differences. For it is the area of human undertaking where the confusion is greatest and its effects the most pernicious. For instance, we often have the mistaken impression that a strategy is an excellent strategy, or an entrepreneur a person endowed with "vision," or a trader a talented trader, only to realize that 99.9% of their past performance is attributable to chance, and chance alone. Ask a profitable investor to explain the reasons for his success; he will offer some deep and convincing interpretation of the results. Frequently, these delusions are intentional and deserve to bear the name "charlatanism."

If there is one cause for this confusion between the left and the right sides of our table, it is our inability to think critically—we may enjoy presenting conjectures as truth. It is our nature. Our mind is not equipped with the adequate machinery to handle

Table P.1 Table of Confusion

Presenting the central distinctions used in the book

General

Luck	Skills
Randomness	Determinism
Probability	Certainty
Belief, conjecture	Knowledge, certitude
Theory	Reality
Anecdote, coincidence	Causality, law
Forecast	Prophecy

Market Performance

Lucky idiot	Skilled investor
Survivorship bias	Market outperformance

Finance

Volatility	Return (or drift)
Stochastic variable	Deterministic variable

Physics and Engineering

Noise	Signal

Literary Criticism

None (literary critics do not seem to have a name for things they do not understand)	Symbol

Philosophy of Science

Epistemic probability	Physical probability
Induction	Deduction
Synthetic proposition	Analytic proposition

General Philosophy

Contingent	Certain
Contingent	Necessary (in the Kripke sense)
Contingent	True in all possible worlds

probabilities; such infirmity even strikes the expert, sometimes just the expert.

The nineteenth-century cartoon character, pot-bellied bourgeois Monsieur Prudhomme, carried around a large sword with a double intent: primarily to defend the Republic against its enemies, and secondarily to attack it should it stray from its course. In the same manner, this book has two purposes: to defend science (as a light beam across the noise of randomness), and to attack the scientist when he strays from his course (most disasters come from the fact that individual scientists do not have an innate understanding of standard error or a clue about critical thinking, and likewise have proved both incapable of dealing with probabilities in the social sciences and incapable of accepting such fact). As a practitioner of uncertainty I have seen more than my share of snake-oil salesmen dressed in the garb of scientists, particularly those operating in economics. The greatest fools of randomness will be found among these.

We are flawed beyond repair, at least for this environment—but it is only bad news for those utopians who believe in an idealized humankind. Current thinking presents the two following polarized visions of man, with little shades in between. On the one hand there is your local college English professor; your great-aunt Irma, who never married and liberally delivers sermons; your how-to-reach-happiness-in-twenty-steps and how-to-become-a-better-person-in-a-week book writer. It is called the Utopian Vision, associated with Rousseau, Godwin, Condorcet, Thomas Paine, and conventional normative economists (of the kind to ask you to make rational choices because that is what is deemed good for you), etc. They believe in reason and rationality—that we should overcome cultural impediments on our way to becoming a better human race—thinking we can control our nature at will and transform it by mere edict in order to attain, among other things, happiness and rationality. Basically this category would include those who think that the cure for obesity is to inform people that they should be healthy.

On the other hand there is the Tragic Vision of humankind that believes in the existence of inherent limitations and flaws in the way we think and act and requires an acknowledgment of this fact as a basis for any individual and collective action. This category of people includes Karl Popper (falsificationism and distrust of intellectual "answers," actually of anyone who is confident that he knows anything with certainty), Friedrich Hayek and Milton Friedman (suspicion of governments), Adam Smith (intention of man), Herbert Simon (bounded rationality), Amos Tversky and Daniel Kahneman (heuristics and biases), the speculator George Soros, etc. The most neglected one is the misunderstood philosopher Charles Sanders Peirce, who was born a hundred years too early (he coined the term scientific "fallibilism" in opposition to Papal infallibility). Needless to say that the ideas of this book fall squarely into the Tragic category: We are faulty and there is no need to bother trying to correct our flaws. We are so defective and so mismatched to our environment that we can just work around these flaws. I am convinced of that after spending almost all my adult and professional years in a fierce fight between my brain (not *Fooled by Randomness*) and my emotions (completely *Fooled by Randomness*) in which the only success I've had is in going around my emotions rather than rationalizing them. Perhaps ridding ourselves of our humanity is not in the works; we need wily tricks, not some grandiose moralizing help. As an empiricist (actually a skeptical empiricist) I despise the moralizers beyond anything on this planet: I still wonder why they blindly believe in ineffectual methods. Delivering advice assumes that our cognitive apparatus rather than our emotional machinery exerts some meaningful control over our actions. We will see how modern behavioral science shows this to be completely untrue.

My colleague Bob Jaeger (he followed the opposite course of

mine of moving from philosophy professor to trader) presents a more potent view of the dichotomy: There are those who think that there are easy clear-cut answers and those who don't think that simplification is possible without severe distortion (his hero: Wittgenstein; his villain: Descartes). I am enamored of the difference as I think that the generator of the *Fooled by Randomness* problem, the false belief in determinism, is also associated with such reduction of the dimensionality of things. As much as you believe in the "keep-it-simple-stupid" it is the *simplification* that is dangerous.

This author hates books that can be easily guessed from the table of contents (not many people read textbooks for pleasure)—but a hint of what comes next seems in order. The book is composed of three parts. The first is an introspection into Solon's warning, as his outburst on rare events became my lifelong motto. In it we meditate on visible and invisible histories and the elusive property of rare events (black swans). The second presents a collection of probability biases I encountered (and suffered from) in my career in randomness—ones that continue to fool me. The third illustrates my personal jousting with my biology and concludes the book with a presentation of a few practical (wax in my ears) and philosophical (stoicism) aids. Before the "enlightenment" and the age of rationality, there was in the culture a collection of tricks to deal with our fallibility and reversals of fortunes. The elders can still help us with some of their ruses.

Part I

•

SOLON'S WARNING

Skewness, Asymmetry, Induction

Croesus, King of Lydia, was considered the richest man of his time. To this day Romance languages use the expression "rich as Croesus" to describe a person of excessive wealth. He was said to be visited by Solon, the Greek legislator known for his dignity, reserve, upright morals, humility, frugality, wisdom, intelligence, and courage. Solon did not display the smallest surprise at the wealth and splendor surrounding his host, nor the tiniest admiration for their owner. Croesus was so irked by the manifest lack of impression on the part of this illustrious visitor that he attempted to extract from him some acknowledgment. He asked him if he had known a happier man than him. Solon cited the life of a man who led a noble existence and died while in battle. Prodded for more, he gave similar examples of heroic but terminated lives, until Croesus, irate, asked him point-blank if he was not to be considered the happiest man of all. Solon answered: "The observation of the numerous misfortunes that attend all conditions forbids us to grow insolent upon our present enjoyments, or to admire a man's happiness that may yet, in course of time, suffer change. For the uncertain future has yet to come, with all variety of future; and him only to whom the divinity has [guaranteed] continued happiness until the end we may call happy."

The modern equivalent has been no less eloquently voiced by the baseball coach Yogi Berra, who seems to have translated Solon's outburst from the pure Attic Greek into no less pure Brooklyn English with "it ain't over until it's over," or, in a less dignified manner, with "it ain't over until the fat lady sings." In addition, aside from his use of the vernacular, the Yogi Berra quote presents an advantage of being true, while the meeting between Croesus and Solon was one of those historical facts that benefited from the imagination of the chroniclers, as it was chronologically impossible for the two men to have been in the same location.

Part I is concerned with the degree to which a situation may yet, in the course of time, suffer change. For we can be tricked by situations involving mostly the activities of the goddess Fortuna— Jupiter's firstborn daughter. Solon was wise enough to get the following point; that which came with the help of luck could be taken away by luck (and often rapidly and unexpectedly at that). The flipside, which deserves to be considered as well (in fact it is even more of our concern), is that things that come with little help from luck are more resistant to randomness. Solon also had the intuition of a problem that has obsessed science for the past three centuries. It is called the problem of induction. I call it in this book the *black swan* or the *rare event*. Solon even understood another linked problem, which I call the *skewness* issue; it does not matter how frequently something succeeds if failure is too costly to bear.

Yet the story of Croesus has another twist. Having lost a battle to the redoubtable Persian king Cyrus, he was about to be burned alive when he called Solon's name and shouted (something like) "Solon, you were right" (again this is legend). Cyrus asked about the nature of such unusual invocations, and he told him about Solon's warning. This impressed Cyrus so much that he decided to spare Croesus' life, as he reflected on the possibilities as far as his own fate was concerned. People were thoughtful at that time.

One

•

IF YOU'RE SO RICH,
WHY AREN'T YOU SO SMART?

An illustration of the effect of randomness on social pecking order and jealousy, through two characters of opposite attitudes. On the concealed rare event. How things in modern life may change rather rapidly, except, perhaps, in dentistry.

NERO TULIP

Hit by Lightning

Nero Tulip became obsessed with trading after witnessing a strange scene one spring day as he was visiting the Chicago Mercantile Exchange. A red convertible Porsche, driven at several times the city speed limit, abruptly stopped in front of the entrance, its tires emitting the sound of pigs being slaughtered. A visibly demented athletic man in his thirties, his face flushed red, emerged and ran up the steps as if he were chased by a tiger. He left the car double-parked, its engine running, provoking an angry fanfare of horns. After a long minute, a bored young man clad in a yellow jacket (yellow was the color reserved for clerks) came down the steps, visibly untroubled by the traffic commotion. He drove the car into the underground parking garage—perfunctorily, as if it were his daily chore.

That day Nero Tulip was hit with what the French call a *coup de foudre*, a sudden intense (and obsessive) infatuation that strikes like lightning. "This is for me!" he screamed enthusiastically—he could not help comparing the life of a trader to the alternative lives that could present themselves to him. Academia conjured up the image of a silent university office with rude secretaries; business, the image of a quiet office staffed with slow thinkers and semislow thinkers who express themselves in full sentences.

Temporary Sanity

Unlike a *coup de foudre*, the infatuation triggered by the Chicago scene has not left him more than a decade and a half after the incident. For Nero swears that no other lawful profession in our times could be as devoid of boredom as that of the trader. Furthermore, although he has not yet practiced the profession of high-sea piracy, he is now convinced that even that occupation would present more dull moments than that of the trader.

Nero could best be described as someone who randomly (and abruptly) swings between the deportment and speech manners of a church historian and the verbally abusive intensity of a Chicago pit trader. He can commit hundreds of millions of dollars in a transaction without a blink or a shadow of a second thought, yet agonize between two appetizers on the menu, changing his mind back and forth and wearing out the most patient of waiters.

Nero holds an undergraduate degree in ancient literature and mathematics from Cambridge University. He enrolled in a Ph.D. program in statistics at the University of Chicago but, after completing the prerequisite coursework, as well as the bulk of his doctoral research, he switched to the philosophy department. He called the switch "a moment of temporary sanity," adding to the consternation of his thesis director, who warned him against philosophers and predicted his return back to the fold. He finished writing his thesis in philosophy. But not the Derrida continental style of incomprehensible philosophy (that is, *incomprehensible* to anyone outside of their ranks, like myself). It was quite the opposite; his thesis was on the methodology of statistical inference in its application to the social sciences. In fact, his thesis was indistinguishable from a thesis in mathematical statistics—it was just a bit more thoughtful (and twice as long).

It is often said that philosophy cannot feed its man—but that was not the reason Nero left. He left because philosophy cannot entertain its man. At first, it started looking futile; he recalled his statistics thesis director's warnings. Then, suddenly, it started to look like work. As he became tired of writing papers on some arcane details of his earlier papers, he gave up the academy. The academic debates bored him to tears, particularly when minute points (invisible to the noninitiated) were at stake. Action was what Nero required. The problem, however, was that he selected the academy in the first place in order to kill what he detected was the flatness and tempered submission of employment life.

After witnessing the scene of the trader chased by a tiger, Nero found a trainee spot on the Chicago Mercantile Exchange, the large exchange where traders transact by shouting and gesticulating frenetically. There he worked for a prestigious (but eccentric) *local*, who trained him in the Chicago style, in return for Nero solving his mathematical equations. The energy in the air proved motivating to Nero. He rapidly graduated to the rank of self-employed trader. Then, when he got tired of standing on his feet in the crowd, and straining his vocal cords, he decided to seek employment "upstairs," that is, trading from a desk. He moved to the New York area and took a position with an investment house.

Nero specialized in quantitative financial products, in which he had an early moment of glory, became famous and in demand.

Many investment houses in New York and London flashed huge guaranteed bonuses at him. Nero spent a couple of years shuttling between the two cities, attending important "meetings" and wearing expensive suits. But soon Nero went into hiding; he rapidly pulled back to anonymity—the Wall Street stardom track did not quite fit his temperament. To stay a "hot trader" requires some organizational ambitions and a power hunger that he feels lucky not to possess. He was only in it for the fun—and his idea of fun does not include administrative and managerial work. He is susceptible to conference room boredom and is incapable of talking to businessmen, particularly the run-of-the-mill variety. Nero is allergic to the vocabulary of business talk, not just on plain aesthetic grounds. Phrases like "game plan," "bottom line," "how to get there from here," "we provide our clients with solutions," "our mission," and other hackneyed expressions that dominate meetings lack both the precision and the coloration that he prefers to hear. Whether people populate silence with hollow sentences, or if such meetings present any true merit, he does not know; at any rate he did not want to be part of it. Indeed Nero's extensive social life includes almost no businesspeople. But unlike me (I can be extremely humiliating when someone rubs me the wrong way with inelegant pompousness), Nero handles himself with gentle aloofness in these circumstances.

So, Nero switched careers to what is called proprietary trading. Traders are set up as independent entities, internal funds with their own allocation of capital. They are left alone to do as they please, provided of course that their results satisfy the executives. The name proprietary comes from the fact that they trade the company's own capital. At the end of the year they receive between 7% and 12% of the profits generated. The proprietary trader has all the benefits of self-employment, and none of the burdens of running the mundane details of his own business. He can work any hours he likes, travel at a whim, and engage in all manner of personal pursuits. It is paradise for an intellectual like Nero who dislikes manual work and values unscheduled meditation. He has been doing that for the past ten years, in the employment of two different trading firms.

Modus Operandi

A word on Nero's methods. He is as conservative a trader as one can be in such a business. In the past he has had good years and less than good years—but virtually no truly "bad" years. Over these years he has slowly built for himself a stable nest egg, thanks to an income ranging between $300,000 and (at the peak) $2.5 million. On average, he manages to accumulate $500,000 a year in after-tax money (from an average income of about $1 million); this goes straight into his savings account. In 1993, he had a bad year and was made to feel uncomfortable in his company. Other traders made out much better, so the capital at his disposal was severely reduced, and he was made to feel undesirable at the institution. He then went to get an identical job, down to an identically de-signed workspace, but in a different firm that was friendlier. In the fall of 1994 the traders who had been competing for the great per-formance award blew up in unison during the worldwide bond market crash that resulted from the random tightening by the Fed-eral Reserve Bank of the United States. They are all currently out of the market, performing a variety of tasks. This business has a high mortality rate.

Why isn't Nero more affluent? Because of his trading style—or perhaps his personality. His risk aversion is extreme. Nero's objec-tive is not to maximize his profits, so much as it is to avoid having this entertaining machine called trading taken away from him. Blowing up would mean returning to the tedium of the university or the nontrading life. Every time his risks increase, he conjures up the image of the quiet hallway at the university, the long mornings at his desk spent in revising a paper, kept awake by bad coffee. No, he does not want to have to face the solemn university library where he was bored to tears. "I am shooting for longevity," he is wont to say.

Nero has seen many traders *blow up*, and does not want to get into that situation. *Blow up* in the lingo has a precise meaning; it does not just mean to lose money; it means to lose more money than one ever expected, to the point of being thrown out of the business (the equivalent of a doctor losing his license to practice or a lawyer being disbarred). Nero rapidly exits trades after a predetermined loss. He never sells "naked options" (a strategy that would leave him exposed to large possible losses). He never puts himself in a situation where he can lose more than, say, $1 million—regardless of the probability of such an event. That amount has always been variable; it depends on his accumulated profits for the year. This risk aversion prevented him from making as much money as the other traders on Wall Street who are often called "Masters of the Universe." The firms he has worked for generally allocate more money to traders with a different style from Nero, like John, whom we will encounter soon.

Nero's temperament is such that he does not mind losing small change. "I love taking small losses," he says. "I just need my winners to be large." In no circumstances does he want to be exposed to those rare events, like panics and sudden crashes, that wipe a trader out in a flash. To the contrary, he wants to benefit from them. When people ask him why he does not hold on to losers, he invariably answers that he was trained by "the most chicken of them all," the Chicago trader Stevo who taught him the business. This is not true; the real reason is his training in probability and his innate skepticism.

There is another reason why Nero is not as rich as others in his situation. His skepticism does not allow him to invest any of his own funds outside of treasury bonds. He therefore missed out on the great bull market. The reason he offers is that it could have turned out to be a bear market and a trap. Nero harbors a deep suspicion that the stock market is some form of an investment scam and cannot bring himself to own a stock. The difference with people around him who were enriched by the stock market was that he was cash-flow rich, but his assets did not inflate at all along with the rest of the world (his treasury bonds hardly changed in value). He contrasts himself with one of those start-up technology companies that were massively cash-flow negative, but for which the hordes developed some infatuation. This allowed the owners to become rich from their stock valuation, and thus dependent on the randomness of the market's election of the winner. The difference with his friends of the investing variety is that he did not depend on the bull market, and, accordingly, does not have to worry about a bear market at all. His net worth is not a function of the investment of his savings—he does not want to depend on his investments, but on his cash earnings, for his enrichment. He takes not an inch of risk with his savings, which he invests in the safest possible vehicles. Treasury bonds are safe; they are issued by the United States government, and governments can hardly go bankrupt since they can freely print their own currency to pay back their obligation.

No Work Ethics

Today, at thirty-nine, after fourteen years in the business, he can consider himself comfortably settled. His personal portfolio contains several million dollars in medium-maturity Treasury bonds, enough to eliminate any worry about the future. What he likes most about proprietary trading is that it requires considerably less time than other high-paying professions; in other words it is perfectly compatible with his non-middle-class work ethic. Trading forces someone to think hard; those who merely work hard generally lose their focus and intellectual energy. In addition, they end up drowning in randomness; work ethics, Nero believes, draw people to focus on noise rather than the signal (the difference we established in Table P.1 on page xliii).

This free time has allowed him to carry on a variety of personal interests. As Nero reads voraciously and spends considerable time in the gym and museums, he cannot have a lawyer's or a doctor's schedule. Nero found the time to go back to the statistics department where he started his doctoral studies and finished the "harder science" doctorate in statistics, by rewriting his thesis in more concise terms. Nero now teaches, once a year, a half-semester seminar called *History of Probabilistic Thinking* in the mathematics department of New York University, a class of great originality that draws excellent graduate students. He has saved enough to be able to maintain his lifestyle in the future and has contingency plans perhaps to retire into writing popular essays of the scientific-literary variety, with themes revolving around probability and *indeterminism*—but only if some event in the future causes the markets to shut down. Nero believes that risk-conscious hard work and discipline can lead someone to achieve a comfortable life with a very high probability. Beyond that, it is all randomness: either by taking enormous (and unconscious) risks, or by being extraordinarily lucky. Mild success can be explainable by skills and labor. Wild success is attributable to variance.

There Are Always Secrets

Nero's probabilistic introspection may have been helped out by a dramatic event in his life—one that he kept to himself. A penetrating observer might detect in Nero a measure of suspicious exuberance, an unnatural drive. For his life is not as crystalline as it may seem. Nero harbors a secret, one that will be discussed in time.

JOHN THE HIGH-YIELD TRADER

Through most of the 1990s, across the street from Nero's house stood John's—a much larger one. John was a high-yield trader, but he was not a trader in the style of Nero. A brief professional conversation with him would have revealed that he presented the intellectual depth and sharpness of mind of an aerobics instructor (though not the physique). A purblind man could have seen that John had been doing markedly better than Nero (or, at least, felt compelled to show it). He parked two top-of-the-line German cars in his driveway (his and hers), in addition to two convertibles (one of which was a collectible Ferrari), while Nero had been driving the same VW Cabriolet for almost a decade—and still does.

The wives of John and Nero were acquaintances, of the health-club type, but Nero's wife felt extremely uncomfortable in the company of John's. She felt that the lady was not merely trying to impress her, but was treating her like someone inferior. While Nero had become inured to the sight of traders getting rich (and trying too hard to become sophisticated by turning into wine collectors and opera lovers), his wife had rarely encountered repressed new wealth—the type of people who have felt the sting of indigence at some point in their lives and want to get even by exhibiting their wares. The only dark side of being a trader, Nero often says, is the sight of money being showered on unprepared people who are suddenly taught that Vivaldi's *Four Seasons* is "refined" music. But it was hard for his spouse to be exposed almost daily to the neighbor who kept boasting of the new decorator they just hired. John and his wife were not the least uncomfortable with the fact that their "library" came with the leather-bound books (her health club reading was limited to *People* magazine but her shelves included a selection of untouched books by dead American authors). She also kept discussing unpronounceable exotic locations where they would repair during their vacations without so much as knowing the smallest thing about the places— she would have been hard put to explain on which continent the Seychelles Islands are located. Nero's wife is all too human; although she kept telling herself that she did not want to be in the shoes of John's wife, she felt as if she had been somewhat swamped in the competition of life. Somehow words and reason became ineffectual in front of an oversized diamond, a monstrous house, and a sports car collection.

An Overpaid Hick

Nero also suffered the same ambiguous feeling toward his neighbors. He was quite contemptuous of John, who represented about everything he is not and does not want to be—but there was the social pressure that was starting to weigh on him. In addition, he too would like to have sampled such excessive wealth. Intellectual contempt does not control personal envy. That house across the street kept getting bigger, with addition after addition—and Nero's discomfort kept apace. While Nero had succeeded beyond his wildest dreams, both personally and intellectually, he was starting to consider himself as having missed a chance somewhere. In the pecking order of Wall Street, the arrival of such types as John had caused him to be a significant trader no longer—but while he used to not care about this, John and his house and his cars had started to gnaw away at him. All would have been well if Nero had not had that stupid large house across the street judging him with a superficial standard every morning. Was it the animal pecking order at play, with John's house size making Nero a beta male? Worse even, John was about five years his junior, and, despite a shorter career, was making at least ten times his income.

When they used to run into each other Nero had a clear feeling that John tried to put him down—with barely detectable but no less potent signs of condescension. Some days John ignored him completely. Had John been a remote character, one Nero could only read about in the papers, the situation would have been different. But there John was in flesh and bones and he was his neighbor. The mistake Nero made was to start talking to him, as the rule of pecking order immediately emerged. Nero tried to soothe his discomfort by recalling the behavior of Swann, the character in Proust's *In Search of Time Lost*, a refined art dealer and man of leisure who was at ease with such men as his personal friend the then Prince of Wales, but acted like he had to prove something in the presence of the middle class. It was much easier for Swann to mix with the aristocratic and well-established set of Guermantes than it was with the social-climbing one of the Verdurins, no doubt because he was far more confident in their presence. Like-wise Nero can exact some form of respect from prestigious and prominent people. He regularly takes long meditative walks in Paris and Venice with an erudite Nobel Prize–caliber scientist (the kind of person who no longer has to prove anything) who actively seeks his conversation. A very famous billionaire speculator calls him regularly to ask him his opinion on the valuation of some de-rivative securities. But there he was obsessively trying to gain the respect of some overpaid hick with a cheap New Jersey "Noo-Joyzy" accent. (Had I been in Nero's shoes I would have paraded some of my scorn to John with the use of body language, but again, Nero is a nice person.)

Clearly, John was not as well educated, well bred, physically fit, or perceived as being as intelligent as Nero—but that was not all; he was not even as street-smart as him! Nero has met true street-smart people in the pits of Chicago who exhibit a rapidity of thinking that he could not detect in John. Nero was convinced that the man was a confident shallow-thinker who had done well because he never made an allowance for his vulnerability. But Nero could not, at times, repress his envy—he wondered whether it was an objective evaluation of John, or if it was his feelings of being slighted that led him to such an assessment of John. Perhaps it was Nero who was not quite the best trader. Maybe if he had pushed himself harder or had sought the right opportunity—instead of "thinking," writing articles and reading complicated papers. Perhaps he should have been involved in the high-yield business, where he would have shined among those shallow-thinkers like John.

So Nero tried to soothe his jealousy by investigating the rules of pecking order. Psychologists have shown that most people prefer to make $70,000 when others around them are making $60,000 than to make $80,000 when others around them are making $90,000. Economics, schmeconomics, it is all pecking order, he thought. No such analysis could prevent him from assessing his condition in an absolute rather than a relative way. With John, Nero felt that, for all his intellectual training, he was just another one of those who would prefer to make less money provided others made even less.

Nero thought that there was at least a hint to support the idea of John being merely lucky—in other words Nero, after all, might not need to move away from his neighbor's starter palazzo. There was hope that John would meet his undoing. For John seemed unaware of one large hidden risk he was taking, the risk of blowup, a risk he could not see because he had too short an experience of the market (but also because he was not thoughtful enough to study history). How could John, with his coarse mind, otherwise be making so much money? This business of junk bonds depends on some knowledge of the "odds," a calculation of the probability of the rare (or random) events. What do such fools know about odds? These traders use "quantitative tools" that give them the odds—and Nero disagrees with the methods used. This high-yield market resembles a nap on a railway track. One afternoon, the surprise train would run you over. You make money every month for a long time, then lose a multiple of your cumulative performance in a few hours. He has seen it with option sellers in 1987, 1989, 1992, and 1998. One day they are taken off the exchange floors, accompanied by oversized security men, and nobody ever sees them again. The big house is simply a loan; John might end up as a luxury car salesman somewhere in New Jersey, selling to the new newly rich who no doubt would feel comfortable in his presence. Nero cannot blow up. His less oversized abode, with its four thousand books, is his own. No market event can take it away from him. Every one of his losses is limited. His trader's dignity will never, never be threatened.

John, for his part, thought of Nero as a loser, and a snobbish overeducated loser at that. Nero was involved in a mature business. He believed that he was way over the hill. "These 'prop' traders are dying," he used to say. "They think they are smarter than everybody else, but they are passé."

THE RED-HOT SUMMER

Finally, in September 1998, Nero was vindicated. One morning while leaving to go to work he saw John in his front yard unusually smoking a cigarette. He was not wearing a business suit. He looked humble; his customary swagger was gone. Nero immediately knew that John had been fired. What he did not suspect was that John also lost almost everything he had. We will see more details of John's losses in Chapter 5.

Nero felt ashamed of his feelings of Schadenfreude, the joy humans can experience upon their rivals' misfortunes. But he could not repress it. Aside from it being unchivalrous, it was said to bring bad luck (Nero is weakly superstitious). But in this case, Nero's merriment did not come from the fact that John went back to his place in life, so much as it was from the fact that Nero's methods, beliefs, and track record had suddenly gained in credibility. Nero would be able to raise public money on his track record precisely because such a thing could not possibly happen to him. A repetition of such an event would pay off massively for him. Part of Nero's elation also came from the fact that he felt proud of his sticking to his strategy for so long, in spite of the pressure to be the alpha male. It was also because he would no longer question his trading style when others were getting rich because they misunderstood the structure of randomness and market cycles.

Serotonin and Randomness

Can we judge the success of people by their raw performance and their personal wealth? Sometimes—but not always. We will see how, at any point in time, a large section of businessmen with outstanding track records will be no better than randomly thrown darts. More curiously, and owing to a peculiar bias, cases will abound of the least-skilled businessmen being the richest. However, they will fail to make an allowance for the role of luck in their performance.

Lucky fools do not bear the slightest suspicion that they may be lucky fools—by definition, they do not know that they belong to such a category. They will act as if they deserved the money. Their strings of successes will inject them with so much serotonin (or some similar substance) that they will even fool themselves about their ability to outperform markets (our hormonal system does not know whether our successes depend on randomness). One can notice it in their posture; a profitable trader will walk upright, dominant style—and will tend to talk more than a losing trader. Scientists found out that serotonin, a neurotransmitter, seems to command a large share of our human behavior. It sets a positive feedback, the virtuous cycle, but, owing to an external kick from randomness, can start a reverse motion and cause a vicious cycle. It has been shown that monkeys injected with serotonin will rise in the pecking order, which in turn causes an increase of the serotonin level in their blood—until the virtuous cycle breaks and starts a vicious one (during the vicious cycle failure will cause one to slide in the pecking order, causing a behavior that will bring about further drops in the pecking order). Likewise, an increase in personal performance (regardless of whether it is caused deterministically or by the agency of Lady Fortuna) induces a rise of serotonin in the subject, itself causing an increase of what is commonly called "leadership" ability. One is "on a roll." Some imperceptible changes in deportment, like an ability to express oneself with serenity and confidence, make the subject look credible—as if he truly deserved the shekels. Randomness will be ruled out as a possible factor in the performance, until it rears its head once again and delivers the kick that will induce the downward spiral.

A word on the display of emotions. Almost no one can conceal his emotions. Behavioral scientists believe that one of the main reasons why people become leaders is not from what skills they seem to possess, but rather from what extremely superficial impression they make on others through hardly perceptible physical signals—what we call today "charisma," for example. The biology of the phenomenon is now well studied under the subject heading "social emotions." Meanwhile some historian will "explain" the success in terms of, perhaps, tactical skills, the right education, or some other theoretical reason seen in hindsight. In addition, there seems to be curious evidence of a link between leadership and a form of psychopathology (the sociopath) that encourages the non-blinking, self-confident, insensitive person to rally followers.

People have often had the bad taste of asking me in a social setting if my day in trading was profitable. If my father were there, he would usually stop them by saying "never ask a man if he is from Sparta: If he were, he would have let you know such an important fact—and if he were not, you could hurt his feelings." Likewise, never ask a trader if he is profitable; you can easily see it in his gesture and gait. People in the profession can easily tell if traders are making or losing money; head traders are quick at identifying an employee who is faring poorly. Their face will seldom reveal much, as people consciously attempt to gain control of their facial expressions. But the way they walk, the way they hold the telephone, and the hesitation in their behavior will not fail to reveal their true disposition. On the morning after John had been fired, he certainly lost much of his serotonin—unless it was another substance that researchers will discover in another decade. One cab driver in Chicago explained to me that he could tell if traders he picked up near the Chicago Board of Trade, a futures exchange, were doing well. "They get all puffed up," he said. I found it interesting (and mysterious) that he could detect it so rapidly. I later got some plausible explanation from evolutionary psychology, which claims that such physical manifestations of one's performance in life, just like an animal's dominant condition, can be used for signaling: It makes the winners seem easily visible, which is efficient in mate selection.

YOUR DENTIST IS RICH, VERY RICH

We close this chapter with a hint on the next discussion of resistance to randomness. Recall that Nero can be considered prosperous but not "very rich" by his day's standards. However, according to some strange accounting measure we will see in the next chapter, he is extremely rich *on the average of lives* he could have led—he takes so little risk in his trading career that there could have been very few disastrous outcomes. The fact that he did not experience John's success was the reason he did not suffer his downfall. He would be therefore wealthy according to this unusual (and probabilistic) method of accounting for wealth. Recall that Nero protects himself from the rare event. Had Nero had to relive his professional life a few million times, very few sample paths would be marred by bad luck—but, owing to his conservatism, very few as well would be affected by extreme good luck. That is, his life in stability would be similar to that of an ecclesiastic clock repairman. Naturally, we are discussing only his professional life, excluding his (sometimes volatile) private one.

Arguably, *in expectation*, a dentist is considerably richer than the rock musician who is driven in a pink Rolls Royce, the speculator who bids up the price of impressionist paintings, or the entrepreneur who collects private jets. For one cannot consider a profession without taking into account the average of the people who enter it, not the sample of those who have succeeded in it. We will examine the point later from the vantage point of the survivorship bias, but here, in Part I, we will look at it with respect to resistance to randomness.

Consider two neighbors, John Doe A, a janitor who won the New Jersey lottery and moved to a wealthy neighborhood, compared to John Doe B, his next-door neighbor of more modest condition who has been drilling teeth eight hours a day over the past thirty-five years. Clearly one can say that, thanks to the dullness of his career, if John Doe B had to relive his life a few thousand times since graduation from dental school, the range of possible outcomes would be rather narrow (assuming he is properly insured). At the best, he would end up drilling the rich teeth of the New York Park Avenue residents, while the worst would show him drilling those of some semideserted town full of trailers in the Catskills. Furthermore, assuming he graduated from a very prestigious teeth-drilling school, the range of outcomes would be even more compressed. As to John Doe A, if he had to relive his life a million times, almost all of them would see him performing janitorial activities (and spending endless dollars on fruitless lottery tickets), and one in a million would see him winning the New Jersey lottery.

The idea of taking into account both the observed and unobserved possible outcomes sounds like lunacy. For most people, probability is about what may happen in the future, not events in the observed past; an event that has already taken place has 100% probability, i.e., certainty. I have discussed the point with many people who platitudinously accuse me of confusing myth and reality. Myths, particularly well-aged ones, as we saw with Solon's warning, can be far more potent (and provide us with more experience) than plain reality.

•

Two

A BIZARRE ACCOUNTING METHOD

On alternative histories, a probabilistic view of the world, intellectual fraud, and the randomness wisdom of a Frenchman with steady bathing habits. How journalists are bred to not understand random series of events. Beware borrowed wisdom: How almost all great ideas concerning random outcomes are against conventional sapience. On the difference between correctness and intelligibility.

ALTERNATIVE HISTORY

I start with the platitude that one cannot judge a performance in any given field (war, politics, medicine, investments) by the results, but by the costs of the alternative (i.e., if history played out in a different way). Such substitute courses of events are called *alternative histories*. Clearly, the quality of a decision cannot be solely judged based on its outcome, but such a point seems to be voiced only by people who fail (those who succeed attribute their success to the quality of their decision). Such opinion—"that I followed the best course"—is what politicians on their way out of office keep telling those members of the press who still listen to them—eliciting the customary commiserating "yes, we know" that makes the sting even worse. And like many platitudes, this one, while being too obvious, is not easy to carry out in practice.

Russian Roulette

One can illustrate the strange concept of alternative histories as follows. Imagine an eccentric (and bored) tycoon offering you $10 million to play Russian roulette, i.e., to put a revolver containing one bullet in the six available chambers to your head and pull the trigger. Each realization would count as one history, for a total of six possible histories of equal probabilities. Five out of these six histories would lead to enrichment; one would lead to a statistic, that is, an obituary with an embarrassing (but certainly original) cause of death. The problem is that only one of the histories is observed in reality; and the winner of $10 million would elicit the admiration and praise of some fatuous journalist (the very same ones who unconditionally admire the Forbes 500 billionaires). Like almost every executive I have encountered during an eighteen-year career on Wall Street (the role of such executives in my view being no more than a judge of results delivered in a random manner), the public observes the external signs of wealth without even having a glimpse at the source (we call such source the *generator*). Consider the possibility that the Russian roulette winner would be used as a role model by his family, friends, and neighbors.

While the remaining five histories are not observable, the wise and thoughtful person could easily make a guess as to their attributes. It requires some thoughtfulness and personal courage. In addition, in time, if the roulette-betting fool keeps playing the game, the bad histories will tend to catch up with him. Thus, if a twenty-five-year-old played Russian roulette, say, once a year, there would be a very slim possibility of his surviving until his fiftieth birthday—but, if there are enough players, say thousands of twenty-five-year-old players, we can expect to see a handful of (extremely rich) survivors (and a very large cemetery). Here I have to admit that the example of Russian roulette is more than intellectual to me. I lost a comrade to this "game" during the Lebanese war, when we were in our teens. But there is more. I discovered that I had more than a shallow interest in literature thanks to the effect of Graham Greene's account of his flirt with such a game; it bore a stronger effect on me than the actual events I had recently witnessed. Greene claimed that he once tried to soothe the dullness of his childhood by pulling the trigger on a revolver—making me shiver at the thought that I had at least a one in six probability of having been without his novels.

The reader can see my unusual notion of alternative accounting: $10 million earned through Russian roulette does not have the same value as $10 million earned through the diligent and artful practice of dentistry. They are the same, can buy the same goods, except that one's dependence on randomness is greater than the other. To an accountant, though, they would be identical; to your next-door neighbor too. Yet, deep down, I cannot help but consider them as qualitatively different. The notion of such alternative accounting has interesting intellectual extensions and lends itself to mathematical formulation, as we will see in the next chapter with our introduction of the Monte Carlo engine. Note that such use of mathematics is only illustrative, aiming at getting the intuition of the point, and should not be interpreted as an engineering issue. In other words, one need not actually compute the alternative histories so much as assess their attributes. Mathematics is not just a "numbers game," it is a way of thinking. We will see that probability is a qualitative subject.

Possible Worlds

Note that these ideas of alternative histories have been covered by separate disciplines in intellectual history, worth presenting quickly because they all seem to converge on the same concept of risk and uncertainty (certainty is something that is likely to take place across the highest number of different alternative histories; uncertainty concerns events that should take place in the lowest number of them).

In philosophy, there has been considerable work on the subject starting with Leibniz' idea of possible worlds. For Leibniz, God's mind included an infinity of possible worlds, of which he selected just one. These nonselected worlds are worlds of possibilities, and the one in which I am breathing and writing these lines is just one of them that happened to have been executed. Philosophers also have a branch of logic that specializes in the matter: whether some property holds *across all possible worlds* or if it holds across a single world—with ramifications into the philosophy of language called *possible worlds semantics* with such authors as Saul Kripke.

In physics, there is the many-world interpretation in quantum mechanics (associated with the works of Hugh Everett in 1957) which considers that the universe branches out treelike at each juncture; what we are living now is only one of these many worlds. Taken at a more extreme level, whenever numerous viable possibilities exist, the world splits into many worlds, one world for each different possibility—causing the proliferation of parallel universes. I am an essayist-trader in one of the parallel universes, plain dust in another.

Finally, in economics: Economists studied (perhaps unwittingly) some of the Leibnizian ideas with the possible "states of nature" pioneered by Kenneth Arrow and Gerard Debreu. This analytical approach to the study of economic uncertainty is called the "state space" method—it happens to be the cornerstone of neoclassical economic theory and mathematical finance. A simplified version is called "scenario analysis," the series of "what-ifs" used in, say, the forecasting of sales for a fertilizer plant under different world conditions and demands for the (smelly) product.

An Even More Vicious Roulette

Reality is far more vicious than Russian roulette. First, it delivers the fatal bullet rather infrequently, like a revolver that would have hundreds, even thousands, of chambers instead of six. After a few dozen tries, one forgets about the existence of a bullet, under a numbing false sense of security. The point is dubbed in this book the *black swan problem*, which we cover in Chapter 7, as it is linked to the problem of induction, a problem that has kept a few thinkers awake at night. It is also related to a problem called *denigration of history*, as gamblers, investors, and decision-makers feel that the sorts of things that happen to others would not necessarily happen to them.

Second, unlike a well-defined, precise game like Russian roulette, where the risks are visible to anyone capable of multiplying and dividing by six, one does not observe the barrel of reality. Very rarely is the generator visible to the naked eye. One is thus capable of unwittingly playing Russian roulette—and calling it by some alternative "low risk" name. We see the wealth being generated, never the processor, a matter that makes people lose sight of their risks, and never consider the losers. The game seems terribly easy and we play along carelessly. Even scientists with all their sophistication in calculating probabilities cannot deliver any meaningful answer about the odds, since knowledge of these depends on our witnessing the barrel of reality—of which we generally know nothing.

Finally, there is an ingratitude factor in warning people about something abstract (by definition anything that did not happen is abstract). Say you engage in a business of protecting investors from rare events by constructing packages that shield them from their sting (something I have done on occasion). Say that nothing happens during the period. Some investors will complain about your spending their money; some will even try to make you feel sorry: "You wasted my money on insurance last year; the factory did not burn, it was a stupid expense. You should only insure for events that happen." One investor came to see me fully expecting me to be apologetic (it did not work). But the world is not that homogeneous: There are some (though very few) who will call you to express their gratitude and thank you for having protected them from the events that did not take place.

SMOOTH PEER RELATIONS

The degree of resistance to randomness in one's life is an abstract idea, part of its logic counterintuitive, and, to confuse matters, its realizations nonobservable. But I have been increasingly devoted to it—for a collection of personal reasons I will leave for later. Clearly my way of judging matters is probabilistic in nature; it relies on the notion of what could have *probably* happened, and requires a certain mental attitude with respect to one's observations. I do not recommend engaging an accountant in a discussion about such probabilistic considerations. For an accountant a number is a number. If he were interested in probability he would have gotten involved in more introspective professions—and would be inclined to make a costly mistake on your tax return.

While we do not see the roulette barrel of reality, some people give it a try; it takes a special mindset to do so. Having seen hundreds of people enter and exit my profession (characterized by extreme dependence on randomness), I have to say that those who have had a modicum of scientific training tend to go the extra mile. For many, such thinking is second nature. This might not necessarily come from their scientific training *per se* (beware of causality), but possibly from the fact that people who have decided at some point in their lives to devote themselves to scientific research tend to have an ingrained intellectual curiosity and a natural tendency for such introspection. Particularly thoughtful are those who had to abandon scientific studies because of their inability to keep focused on a narrowly defined problem (or, in Nero's case, the minute arcane details and petty arguments). Without excessive intellectual curiosity it is almost impossible to complete a Ph.D. thesis these days; but without a desire to narrowly specialize, it is impossible to make a scientific career. (There is a distinction, however, between the mind of a pure mathematician thriving on abstraction and that of a scientist consumed by curiosity. A mathematician is absorbed in what goes into his head while a scientist searches into what is outside of himself.) However, some people's concern for randomness can be excessive; I have even seen people trained in some fields, like, say, quantum mechanics, push the idea to the other extreme, only seeing alternative histories (in the many-world interpretation) and ignoring the one that actually took place.

Some traders can be unexpectedly introspective about randomness. Not long ago I had dinner at the bar of a Tribeca restaurant with Lauren Rose, a trader who was reading an early draft of this book. We flipped a coin to see who was going to pay for the meal. I lost and paid. He was about to thank me when he abruptly stopped and said that he paid for half of it *probabilistically*.

I thus view people distributed across two polar categories: On one extreme, those who never accept the notion of randomness; on the other, those who are tortured by it. When I started on Wall Street in the 1980s, trading rooms were populated with people with a "business orientation," that is, generally devoid of any introspection, flat as a pancake, and likely to be fooled by randomness. Their failure rate was extremely high, particularly when financial instruments gained in complexity. Somehow, tricky products, like exotic options, were introduced and carried counterintuitive payoffs that were too difficult for someone of such culture to handle. They dropped like flies; I do not think that many of the hundreds of MBAs of my generation I met on Wall Street in the 1980s still engage in such forms of professional and disciplined risk taking.

Salvation via Aeroflot

The 1990s witnessed the arrival of people of richer and more interesting backgrounds, which made the trading rooms far more entertaining. I was saved from the conversation of MBAs. Many scientists, some of them extremely successful in their field, arrived with a desire to make a buck. They, in turn, hired people who resembled them. While most of these people were not Ph.D.s (indeed, the Ph.D. is still a minority), the culture and values suddenly changed, becoming more tolerant of intellectual depth. It caused an increase in the already high demand for scientists on Wall Street, owing to the rapid development of financial instruments. The dominant specialty was physics, but one could find all manner of quantitative backgrounds among them. Russian, French, Chinese, and Indian accents (by order) began dominating in both New York and London. It was said that every plane from Moscow had at least its back row full of Russian mathematical physicists en route to Wall Street (they lacked the street smarts to get good seats). One could hire very cheap labor by going to JFK airport with a (mandatory) translator, randomly interviewing those who fit the stereotype. Indeed, by the late 1990s one could get someone trained by a world-class scientist for almost half the price of an MBA. As they say, marketing is everything; these guys do not know how to sell themselves.

I had a strong bias in favor of Russian scientists; many can be put to active use as chess coaches (I also got a piano teacher out of the process). In addition, they are extremely helpful in the interview process. When MBAs apply for trading positions, they frequently boast "advanced" chess skills on their résumés. I recall the MBA career counselor at Wharton recommending our advertising chess skills "because it sounds intelligent and strategic." MBAs, typically, can interpret their superficial knowledge of the rules of the game into "expertise." We used to verify the accuracy of claims of chess expertise (and the character of the applicant) by pulling a chess set out of a drawer and telling the student, now turning pale: "Yuri will have a word with you."

The failure rate of these scientists, though, was better, but only slightly so than that of MBAs; but it came from another reason, linked to their being on average (but only on average) devoid of the smallest bit of practical intelligence. Some successful scientists had the judgment (and social graces) of a doorknob—but by no means all of them. Many people were capable of the most complex calculations with utmost rigor when it came to equations, but were totally incapable of solving a problem with the smallest connection to reality; it was as if they understood the letter but not the spirit of the math (we will see more on such dual thinking with the two systems of reasoning problem in Chapter 11). I am convinced that X, a likeable Russian man of my acquaintance, has two brains: one for math and another, considerably inferior one, for everything else (which included solving problems related to the mathematics of finance). But on occasion a fast-thinking scientific-minded person with street smarts would emerge. Whatever the benefits of such population shift, it improved our chess skills and provided us with quality conversation during lunchtime—it extended the lunch hour considerably. Consider that I had in the 1980s to chat with colleagues who had an MBA or tax accounting background and were capable of the heroic feat of discussing FASB standards. I have to say that their interests were not too contagious. The interesting thing about these physicists did not lie in their ability to discuss fluid dynamics; it is that they were naturally interested in a variety of intellectual subjects and provided pleasant conversation.

Solon Visits Regine's Nightclub

As the reader may already suspect, my opinions about randomness have not earned me the smoothest of relations with some of my peers during my Wall Street career (many of whom the reader can see indirectly—but only indirectly—portrayed in these chapters). But where I had uneven relations was with some of those who had the misfortune of being my bosses. For I had two bosses in my life of contrasting characteristics in about every trait.

The first, whom I will call Kenny, was the epitome of the sub- urban family man. He would be of the type to coach soccer on Saturday morning, and invite his brother-in-law for a Sunday af- ternoon barbecue. He gave the appearance of someone I would trust with my savings—indeed he rose quite rapidly in the in- stitution in spite of his lack of technical competence in financial derivatives (his firm's claim to fame). But he was too much a no- nonsense person to make out my logic. He once blamed me for not being impressed with the successes of some of his traders who did well during the bull market for European bonds of 1993, whom I openly considered nothing better than random gun- slingers. I tried presenting him with the notion of survivorship bias (Part II of this book) in vain. His traders have all exited the busi- ness since then "to pursue other interests" (including him). But he gave the appearance of being a calm, measured man, who spoke his mind and knew how to put the other person at ease during a conversation. He was articulate, extremely presentable thanks to his athletic looks, well measured in his speech, and endowed with the extremely rare quality of being an excellent listener. His per- sonal charm allowed him to win the confidence of the chairman— but I could not conceal my disrespect, particularly as he could not make out the nature of my conversation. In spite of his conserva- tive looks he was a perfect time bomb, ticking away.

The second, whom I will call Jean-Patrice, in contrast, was a moody Frenchman with an explosive temper and a hyperaggres- sive personality. Except for those he truly liked (not that many), he was expert at making his subordinates uncomfortable, putting them in a state of constant anxiety. He greatly contributed to my formation as a risk taker; he is one of the very rare people who have the guts to care only about the generator, entirely oblivious of the results. He presented the wisdom of Solon, but, while one would expect someone with such personal wisdom and such understanding of randomness to lead a dull life, he lived a colorful one. In contrast with Kenny, who wore conservative dark suits and white shirts (his only indulgence was flashy equestrian Hermès ties), Jean-Patrice dressed like a peacock: blue shirts, plaid sports coats stuffed with gaudy silk pocket squares. No family-minded man, he rarely came to work before noon—though I can safely say that he carried his work with him to the most unlikely places. He frequently called me from *Regine's*, an upscale nightclub in New York, waking me up at three in the morning to discuss some small (and irrelevant) details of my risk exposure. In spite of his slight corpulence, women seemed to find him irresistible; he frequently disappeared at midday and was unreachable for hours. His advantage might have been in his being a New York Frenchman with steady bathing habits. Once he invited me to discuss an urgent business issue with him. Characteristically, I found him mid-afternoon in a strange "club" in Paris that carried no nameplate and where he sat with documents strewn across the table from him. Sipping champagne, he was simultaneously caressed by two scantily dressed young ladies. Strangely, he involved them in the conversation as if they were part of the meeting. He even had one of the ladies pick up his constantly ringing mobile phone as he did not want our conversation to be interrupted.

I am still amazed at this flamboyant man's obsession with risks, which he constantly played in his head—he literally thought of everything that could possibly happen. He forced me to make an alternative plan should a plane crash into the office building (way before the events of September 2001)—and fumed at my answer that the financial condition of his department would be of small interest to me in such circumstances. He had a horrible reputation as a philanderer, a temperamental boss capable of firing someone at a whim, yet he listened to me and understood every word I had to say, encouraging me to go the extra mile in my study of randomness. He taught me to look for the invisible risks of blowup in any portfolio. Not coincidentally, he has an immense respect for science and an almost fawning deference for scientists; a decade or so after we worked together he showed up unexpectedly during the defense of my doctoral thesis, smiling from the back of the room. While Kenny knew how to climb the ladder of an institution, reaching a high level in the organization before being forced out, Jean-Patrice did not have such a happy career, a matter that taught me to beware of mature financial institutions.

It can be disturbing for many self-styled "bottom line"–oriented people to be questioned about the histories that did not take place rather than the ones that actually happened. Clearly, to a no-nonsense person of the "successful in business" variety, my language (and, I have to reckon, some traits of my personality) appears strange and incomprehensible. To my amusement, the argument appears offensive to many.

The contrast between Kenny and Jean-Patrice is not a mere coincidence that I happened to witness in a protracted career. Beware the spendthrift "businesswise" person; the cemetery of markets is disproportionately well stocked with the self-styled "bottom line" people. In contrast with their customary Masters of the Universe demeanor, they suddenly look pale, humble, and hormone-deprived on the way to the personnel office for the customary discussion of the severance agreement.

GEORGE WILL IS NO SOLON: ON COUNTERINTUITIVE TRUTHS

Realism can be punishing. Probabilistic skepticism is worse. It is difficult to go about life wearing probabilistic glasses, as one starts seeing fools of randomness all around, in a variety of situations—

obdurate in their perceptional illusion. To start, it is impossible to read a historian's analysis without questioning the inferences: We know that Hannibal and Hitler were mad in their pursuits, as Rome is not today Phoenician-speaking and Times Square in New York currently exhibits no swastikas. But what of all those generals who were equally foolish, but ended up winning the war and consequently the esteem of the historical chronicler? It is hard to think of Alexander the Great or Julius Caesar as men who won only in the visible history, but who could have suffered defeat in others. If we have heard of them, it is simply because they took considerable risks, along with thousands of others, and happened to win. They were intelligent, courageous, noble (at times), had the highest possible obtainable culture in their day—but so did thousands of others who live in the musty footnotes of history. Again I am not contesting that they won their wars—only the claims concerning the quality of their strategies. (My very first impression upon a recent rereading of the *Iliad*, the first in my adulthood, is that the epic poet did not judge his heroes by the result: Heroes won and lost battles in a manner that was totally independent of their own valor; their fate depended upon totally external forces, generally the explicit agency of the scheming gods (not devoid of nepotism). Heroes are heroes because they are heroic in behavior, not because they won or lost. Patrocles does not strike us as a hero because of his accomplishments (he was rapidly killed) but because he preferred to die than see Achilles sulking into inaction. Clearly, the epic poets understood invisible histories. Also later thinkers and poets had more elaborate methods for dealing with randomness, as we will see with stoicism.

Listening to the media, mostly because I am not used to it, can cause me on occasion to jump out of my seat and become emotional in front of the moving image (I grew up with no television and was in my late twenties when I learned to operate a TV set). One illustration of a dangerous refusal to consider alternative histories is provided by the interview that media person George Will, a "commentator" of the extensively commenting variety, conducted with Professor Robert Shiller, a man known to the public for his bestselling book *Irrational Exuberance*, but known to the connoisseur for his remarkable insights about the structure of market randomness and volatility (expressed in the precision of mathematics).

The interview is illustrative of the destructive aspect of the media, in catering to our heavily warped common sense and biases. I was told that George Will was very famous and extremely respected (that is, for a journalist). He might even be someone of the utmost intellectual integrity; his profession, however, is merely to sound smart and intelligent to the hordes. Shiller, on the other hand, understands the ins and outs of randomness; he is trained to deal with rigorous argumentation, but does sound less smart in public because his subject matter is highly counterintuitive. Shiller had been pronouncing the stock market to be overpriced for a long time. George Will indicated to Shiller that had people listened to him in the past they would have lost money, as the market has more than doubled since he started pronouncing it overvalued. To such a journalistic and well-sounding (but senseless) argument, Shiller was unable to respond except to explain that the fact that he was wrong in one single market call should not carry undue significance. Shiller, as a scientist, did not claim to be a prophet or one of the entertainers who comment on the markets on the evening news. Yogi Berra would have had a better time with his confident comment on the fat lady not having sung yet.

I could not understand what Shiller, untrained to compress his ideas into vapid sound bites, was doing on such a TV show. Clearly, it is foolish to think that an irrational market cannot become even more irrational; Shiller's views on the rationality of the market are not invalidated by the argument that he was wrong in the past. Here I could not help seeing in the person of George Will the representative of so many nightmares in my career; my attempting to prevent someone from playing Russian roulette for $10 million and seeing journalist George Will humiliating me in public by saying that had the person listened to me it would have cost him a considerable fortune. In addition, Will's comment was not an off-the-cuff remark; he wrote an article on the matter discussing Shiller's bad "prophecy." Such tendency to make and unmake prophets based on the fate of the roulette wheel is symptomatic of our ingrained inability to cope with the complex structure of randomness prevailing in the modern world. Mixing forecast and prophecy is symptomatic of randomness-foolishness (prophecy belongs to the right column; forecast is its mere left-column equivalent).

Humiliated in Debates

Clearly, this idea of alternative history does not make intuitive sense, which is where the fun begins. For starters, we are not wired in a way to understand probability, a point that we will examine backward and forward in this book. I will just say at this point that researchers of the brain believe that mathematical truths make little sense to our mind, particularly when it comes to the examination of random outcomes. Most results in probability are entirely counterintuitive; we will see plenty of them. Then why argue with a mere journalist whose paycheck comes from playing on the conventional wisdom of the hordes? I recall that every time I have been humiliated in a public discussion on markets by someone (of the George Will variety) who seemed to present more palatable and easier-to-understand arguments, I turned out (much later) to be right. I do not dispute that arguments should be simplified to their maximum potential; but people often confuse complex ideas that cannot be simplified into a media-friendly statement as symptomatic of a confused mind. MBAs learn the concept of clarity and simplicity—the five-minute-manager take on things. The concept may apply to the business plan for a fertilizer plant, but not to highly probabilistic arguments—which is the reason I have anecdotal evidence in my business that MBAs tend to blow up in financial markets, as they are trained to simplify matters a couple of steps beyond their requirement. (I beg the MBA reader not to take offense; I am myself the unhappy holder of the degree.)

A Different Kind of Earthquake

Try the following experiment. Go to the airport and ask travelers en route to some remote destination how much they would pay for an insurance policy paying, say, a million tugrits (the currency of Mongolia) if they died during the trip (for any reason). Then ask another collection of travelers how much they would pay for insurance that pays the same in the event of death from a terrorist act (and only a terrorist act). Guess which one would command a higher price? Odds are that people would rather pay for the second policy (although the former includes death from terrorism). The psychologists Daniel Kahneman and Amos Tversky figured this out several decades ago. The irony is that one of the sampled populations did not include people on the street, but professional predictors attending some society of forecasters' annual meeting. In a now famous experiment they found that the majority of people, whether predictors or nonpredictors, will judge a deadly flood (causing thousands of deaths) caused by a California earthquake to be more likely than a fatal flood (causing thousands of deaths) occurring somewhere in North America (which happens to include California). As a derivatives trader I noticed that people do not like to insure against something abstract; the risk that merits their attention is always something vivid.

This brings us to a more dangerous dimension of journalism. We just saw how the scientifically hideous George Will and his colleagues can twist arguments to sound right without being right. But there is a more general impact by information providers in biasing the representation of the world one gets from the delivered information. It is a fact that our brain tends to go for superficial clues when it comes to risk and probability, these clues being largely determined by what emotions they elicit or the ease with which they come to mind. In addition to such problems with the perception of risk, it is also a scientific fact, and a shocking one, that both risk detection and risk avoidance are not mediated in the "thinking" part of the brain but largely in the emotional one (the "risk as feelings" theory). The consequences are not trivial: It means that rational thinking has little, very little, to do with risk avoidance. Much of what rational thinking seems to do is rationalize one's actions by fitting some logic to them.

In that sense the description coming from journalism is certainly not just an unrealistic representation of the world but rather the one that can fool you the most by grabbing your attention via your emotional apparatus—the *cheapest to deliver* sensation. Take the mad cow "threat" for example: Over a decade of hype, it only killed people (in the highest estimates) in the hundreds as compared to car accidents (several hundred thousands!)—except that the journalistic description of the latter would not be commercially fruitful. (Note that the risk of dying from food poisoning or in a car accident on the way to a restaurant is greater than dying from mad cow disease.) This sensationalism can divert empathy toward wrong causes: cancer and malnutrition being the ones that suffer the most from the lack of such attention. Malnutrition in Africa and Southeast Asia no longer causes the emotional impact—so it literally dropped out of the picture. In that sense the mental probabilistic map in one's mind is so geared toward the sensational that one would realize informational gains by dispensing with the news. Another example concerns the volatility of markets. In people's minds lower prices are far more "volatile" than sharply higher moves. In addition, volatility seems to be determined not by the actual moves but by the tone of the media.

The market movements in the eighteen months after September 11, 2001, were far smaller than the ones that we faced in the eighteen months prior—but somehow in the mind of investors they were very volatile. The discussions in the media of the "terrorist threats" magnified the effect of these market moves in people's heads. This is one of the many reasons that journalism may be the greatest plague we face today—as the world becomes more and more complicated and our minds are trained for more and more simplification.

Proverbs Galore

Beware the confusion between correctness and intelligibility. Part of conventional wisdom favors things that can be explained rather instantly and "in a nutshell"—in many circles it is considered law. Having attended a French elementary school, a *lycée primaire*, I was trained to rehash Boileau's adage:

> *Ce qui se conçoit bien s'énonce clairement*
> *Et les mots pour le dire viennent aisément*

What is easy to conceive is clear to express / Words to say it would come effortlessly.

The reader can imagine my disappointment at realizing, while growing up as a practitioner of randomness, that most poetic sounding adages are plain wrong. Borrowed wisdom can be vicious. I need to make a huge effort not to be swayed by well-sounding remarks. I remind myself of Einstein's remark that common sense is nothing but a collection of misconceptions acquired by age eighteen. Furthermore, *What sounds intelligent in a conversation or a meeting, or, particularly, in the media, is suspicious.*

Any reading of the history of science would show that almost all the smart things that have been proven by science appeared like lunacies at the time they were first discovered. Try to explain to a *Times* (of London) journalist in 1905 that time slows down when one travels (even the Nobel committee never granted Einstein the prize on account of his insight on special relativity). Or to someone with no exposure to physics that there are places in our universe where time does not exist. Try to explain to Kenny that, although his star trader "proved" to be extremely successful, I have enough arguments to convince him that he is a dangerous idiot.

Risk Managers

Corporations and financial institutions have recently created the strange position of risk manager, someone who is supposed to monitor the institution and verify that it is not too deeply involved in the business of playing Russian roulette. Clearly, having been burned a few times, the incentive is there to have someone take a look at the generator, the roulette that produces the profits and losses. Although it is more fun to trade, many extremely smart people among my friends (including Jean-Patrice) felt attracted by such positions. It is an important and attractive fact that the average risk manager earns more than the average trader (particularly when we take into account the number of traders thrown out of the business: While a ten-year survival rate for a trader is in the single digits, that of a risk manager is close to 100%). "Traders come and go; risk managers are here to stay." I keep thinking of taking such a position both on economic grounds (as it is probabilistically more profitable) and because the job offers more intellectual content than the one consisting in just buying and selling, and allows one to integrate research and execution. Finally, a risk manager's blood has smaller quantities of the harmful kind of stress hormones. But something has held me back, aside from the irrationality of wanting the pains and entertainment from the emotions of speculation. The risk managers' job feels strange: As we said, the generator of reality is not observable. They are limited in their power to stop profitable traders from taking risks, given that they would, *ex post*, be accused by the George Wills around of costing the shareholder some precious opportunity shekels. On the other hand, the occurrence of a blowup would cause them to be responsible for it. What to do in such circumstances?

Their focus becomes to play politics, cover themselves by issuing vaguely phrased internal memoranda that warn against risk-taking activities yet stop short of completely condemning it, lest they lose their job. Like a doctor torn between the two types of errors, the false positive (telling the patient he has cancer when in fact he does not) and the false negative (telling the patient he is healthy when in fact he has cancer), they need to balance their existence with the fact that they inherently need some margin of error in their business.

Epiphenomena

From the standpoint of an institution, the existence of a risk manager has less to do with actual risk reduction than it has to do with the *impression* of risk reduction. Philosophers since Hume and modern psychologists have been studying the concept of epiphenomenalism, or when one has the illusion of cause-and-effect. Does the compass move the boat? By "watching" your risks, are you effectively reducing them or are you giving yourself the feeling that you are doing your duty? Are you like a chief executive officer or just an observing press officer? Is such illusion of control harmful?

I conclude the chapter with a presentation of the central paradox of my career in financial randomness. By definition, I go against the grain, so it should come as no surprise that my style and methods are neither popular nor easy to understand. But I have a dilemma: On the one hand, I work with others in the real world, and the real world is not just populated with babbling but ultimately inconsequential journalists. So my wish is for people in general to remain fools of randomness (so I can trade against them), yet for there to remain a minority intelligent enough to value my methods and hire my services. In other words, I need people to remain fools of randomness, but not all of them. I was fortunate to meet Donald Sussman, who corresponds to such an ideal partner; he helped me in the second stage of my career by freeing me from the ills of employment. My greatest risk is to become successful, as it would mean that my business is about to disappear; strange business, ours.

•

Three

A MATHEMATICAL MEDITATION ON HISTORY

On Monte Carlo simulation as a metaphor for understand-ing a sequence of random historical events. On randomness and artificial history. Age is beauty, almost always, and the new and the young are generally toxic. Send your history pro-fessor to an introductory class on sampling theory.

Europlayboy Mathematics

The stereotype of a pure mathematician presents an anemic man with a shaggy beard and grimy and uncut fingernails silently laboring on a Spartan but disorganized desk. With thin shoulders and a pot belly, he sits in a grubby office, totally ab-sorbed in his work, oblivious to the grunginess of his surroundings. He grew up in a communist regime and speaks English with an as-tringent and throaty Eastern European accent. When he eats, crumbs of food accumulate in his beard. With time he becomes more and more absorbed in his subject matter of pure theorems, reaching levels of ever increasing abstraction. The American public was recently exposed to one of these characters with the Unabomber, the bearded and recluse mathematician who lived in a hut and took to murdering people who promoted modern technology. No journalist was capable of even coming close to describing the subject matter of his thesis, "Complex Boundaries," as it has no intelligible equivalent—a complex number being an entirely abstract and imaginary number that includes the square root of minus one, an object that has no analog outside of the world of mathematics.

The name Monte Carlo conjures up the image of a suntanned urbane man of the Europlayboy variety entering a casino under a whiff of the Mediterranean breeze. He is an apt skier and tennis player, but also can hold his own in chess and bridge. He drives a gray sports car, dresses in a well-ironed Italian handmade suit, and speaks carefully and smoothly about mundane, but real, matters, those a journalist can easily describe to the public in compact sentences. Inside the casino he astutely counts the cards, mastering the odds, and bets in a studied manner, his mind producing precise calculations of his optimal betting size. He could be James Bond's smarter lost brother.

Now when I think of Monte Carlo mathematics, I think of a happy combination of the two: The Monte Carlo man's realism without the shallowness, combined with the mathematician's intuitions without the excessive abstraction. For indeed this branch of mathematics is of immense practical use—it does not present the same dryness commonly associated with mathematics. I became addicted to it the minute I became a trader. It shaped my thinking in most matters related to randomness. Most of the examples used in this book were created with my Monte Carlo generator, which I introduce in this chapter. Yet it is far more a way of thinking than a computational method. Mathematics is principally a tool to meditate, rather than to compute.

The Tools

The notion of alternative histories discussed in the last chapter can be extended considerably and subjected to all manner of technical refinement. This brings us to the tools used in my profession to toy with uncertainty. I will outline them next. Monte Carlo methods, in brief, consist of creating artificial history using the following concepts.

First, consider the sample path. The invisible histories have a scientific name, *alternative sample paths*, a name borrowed from the field of mathematics of probability called stochastic processes. The notion of path, as opposed to outcome, indicates that it is not a mere MBA-style scenario analysis, but the examination of a sequence of scenarios along the course of time. We are not just concerned with where a bird can end up tomorrow night, but rather with all the various places it can possibly visit during the time interval. We are not concerned with what the investor's worth would be in, say, a year, but rather of the heart-wrenching rides he may experience during that period. The word *sample* stresses that one sees only one realization among a collection of possible ones. Now, a sample path can be either deterministic or random, which brings the next distinction.

A *random sample path*, also called a random run, is the mathematical name for such a succession of virtual historical events, starting at a given date and ending at another, except that they are subjected to some varying level of uncertainty. However, the word *random* should not be mistaken for equiprobable (i.e., having the same probability). Some outcomes will give a higher probability than others. An example of a random sample path can be the body temperature of your explorer cousin during his latest bout with typhoid fever, measured hourly from the beginning to the end of his episode. It can also be a simulation of the price of your favorite technology stock, measured daily at the close of the market, over, say, one year. Starting at $100, in one scenario it can end up at $20 having seen a high of $220; in another it can end up at $145 having seen a low of $10. Another example is the evolution of your wealth during an evening at a casino. You start with $1,000 in your pocket, and measure it every fifteen minutes. In one sample path

you have $2,200 at midnight; in another you barely have $20 left
for a cab fare.

Stochastic processes refer to the dynamics of events unfolding
with the course of time. Stochastic is a fancy Greek name for random. This branch of probability concerns itself with the study of
the evolution of successive random events—one could call it the
mathematics of history. The key about a process is that it has time
in it.

What is a Monte Carlo generator? Imagine that you can replicate a perfect roulette wheel in your attic without having recourse
to a carpenter. Computer programs can be written to simulate just
about anything. They are even better (and cheaper) than the
roulette wheel built by your carpenter, as this physical version
may be inclined to favor one number more than others owing to a
possible slant in its build or the floor of your attic. These are called
the biases.

Monte Carlo simulations are closer to a toy than anything I have
seen in my adult life. One can generate thousands, perhaps millions, of random sample paths, and look at the prevalent characteristics of some of their features. The assistance of the computer is
instrumental in such studies. The glamorous reference to Monte
Carlo indicates the metaphor of simulating the random events in
the manner of a virtual casino. One sets conditions believed to resemble the ones that prevail in reality, and launches a collection of
simulations around possible events. With no mathematical literacy
we can launch a Monte Carlo simulation of an eighteen-year-old
Christian Lebanese successively playing Russian roulette for a
given sum, and see how many of these attempts result in enrichment, or how long it takes on average before he hits the obituary.
We can change the barrel to contain 500 holes, a matter that
would decrease the probability of death, and see the results.

Monte Carlo simulation methods were pioneered in martial
physics in the Los Alamos laboratory during the A-bomb prepara-

tion. They became popular in financial mathematics in the 1980s, particularly in the theories of the random walk of asset prices. Clearly, we have to say that the example of Russian roulette does not need such apparatus, but many problems, particularly those resembling real-life situations, require the potency of a Monte Carlo simulator.

Monte Carlo Mathematics

It is a fact that "true" mathematicians do not like Monte Carlo methods. They believe that they rob us of the finesse and elegance of mathematics. They call it "brute force." For we can replace a large portion of mathematical knowledge with a Monte Carlo simulator (and other computational tricks). For instance, someone with no formal knowledge of geometry can compute the mysterious, almost mystical Pi. How? By drawing a circle inside of a square, and "shooting" random bullets into the picture (as in an arcade), specifying equal probabilities of hitting any point on the map (something called a uniform distribution). The ratio of bullets inside the circle divided by those inside and outside the circle will deliver a multiple of the mystical Pi, with possibly infinite precision. Clearly, this is not an efficient use of a computer as Pi can be computed analytically, that is, in a mathematical form, but the method can give some users more intuition about the subject matter than lines of equations. Some people's brains and intuitions are oriented in such a way that they are more capable of getting a point in such a manner (I count myself one of those). The computer might not be natural to our human brain; neither is mathematics.

I am not a "native" mathematician, that is, I am someone who does not speak mathematics as a native language, but someone who speaks it with a trace of a foreign accent. For I am not interested in mathematical properties *per se*, only in the application, while a mathematician would be interested in improving mathematics (via theorems and proofs). I proved incapable of concentrating on deciphering a single equation unless I was motivated by a real problem (with a modicum of greed); thus most of what I know comes from derivatives trading—options pushed me to study the math of probability. Many compulsive gamblers, who otherwise would be of middling intelligence, acquire remarkable card-counting skills thanks to their passionate greed.

Another analogy would be with grammar; mathematics is often tedious and insightless grammar. There are those who are interested in grammar for grammar's sake, and those interested in avoiding solecisms while writing documents. Those of us in the second category are called "quants"—like physicists, we have more interest in the employment of the mathematical tool than in the tool itself. Mathematicians are born, never made. Physicists and quants too. I do not care about the "elegance" and "quality" of the mathematics I use so long as I can get the point right. I have recourse to Monte Carlo machines whenever I can. They can get the work done. They are also far more pedagogical, and I will use them in this book for the examples.

Indeed, probability is an introspective field of inquiry, as it affects more than one science, particularly the mother of all sciences: that of knowledge. It is impossible to assess the quality of the knowledge we are gathering without allowing a share of randomness in the manner it is obtained and cleaning the argument from the chance coincidence that could have seeped into its construction. In science, probability and information are treated in exactly the same manner. Literally every great thinker has dabbled with it, most of them obsessively. The two greatest minds to me, Einstein and Keynes, both started their intellectual journeys with it. Einstein wrote a major paper in 1905, in which he was almost the first to examine in probabilistic terms the succession of random events, namely the evolution of suspended particles in a stationary liquid. His article on the theory of the Brownian movement can be used as the backbone of the random walk approach used in financial modeling. As for Keynes, to the literate person he is not the political economist that tweed-clad leftists love to quote, but the author of the magisterial, introspective, and potent *Treatise on Probability*. For before his venturing into the murky field of political economy, Keynes was a probabilist. He also had other interesting attributes (he blew up trading his account after experiencing excessive opulence—people's understanding of probability does not translate into their behavior).

The reader can guess that the next step from such probabilistic introspection is to get drawn into philosophy, particularly the branch of philosophy that concerns itself with knowledge, called epistemology or methodology, or philosophy of science. We will not get into the topic until later in the book.

FUN IN MY ATTIC

Making History

In the early 1990s, like many of my friends in quantitative finance, I became addicted to the various Monte Carlo engines, which I taught myself to build, thrilled to feel that I was generating history, a *Demiurgus*. It can be electrifying to generate virtual histories and watch the dispersion between the various results. Such dispersion is indicative of the degree of resistance to randomness. This is where I am convinced that I have been extremely lucky in my choice of career: One of the attractive aspects of my profession as a quantitative option trader is that I have close to 95% of my day free to think, read, and research (or "reflect" in the gym, on ski slopes, or, more effectively, on a park bench). I also had the privilege of frequently "working" from my well-equipped attic.

The dividend of the computer revolution to us did not come in the flooding of self-perpetuating e-mail messages and access to chat rooms; it was in the sudden availability of fast processors capable of generating a million sample paths per minute. Recall that I never considered myself better than an unenthusiastic equation solver and was rarely capable of prowess in the matter—being better at setting up equations than solving them. Suddenly, my engine allowed me to solve with minimal effort the most intractable of equations. Few solutions became out of reach.

Zorglubs Crowding the Attic

My Monte Carlo engine took me on a few interesting adventures. While my colleagues were immersed in news stories, central bank announcements, earnings reports, economic forecasts, sports results, and, not least, office politics, I started toying with it in fields bordering my home base of financial probability. A natural field of expansion for the amateur is evolutionary biology—the universality of its message and its application to markets are appealing. I started simulating populations of fast-mutating animals called Zorglubs under climatic changes and witnessing the most unexpected of conclusions—some of the results are recycled in Chapter 5. My aim, as a pure amateur fleeing the boredom of business life, was merely to develop intuitions for these events—the sort of intuitions that amateurs build away from the overly detailed sophistication of the professional researcher. I also toyed with molecular biology, generating randomly occurring cancer cells and witnessing some surprising aspects of their evolution. Naturally the analog to fabricating populations of Zorglubs was to simulate a population of "idiotic bull," "impetuous bear," and "cautious" traders under different market regimes, say booms and busts, and to examine their short-term and long-term survival. Under such a structure, "idiotic bull" traders who get rich from the rally would use the proceeds to buy more assets, driving prices higher, until their ultimate shellacking. Bearish traders, though, rarely made it in the boom to get to the bust. My models showed that ultimately almost nobody really survived; bears dropped out like flies in the rally and bulls ended up being slaughtered, as paper profits vanished when the music stopped. But there was one exception; some of those who traded options (I called them option buyers) had remarkable staying power and I wanted to be one of those. How? Because they could buy the insurance against blowup; they could get anxiety-free sleep at night, thanks to the knowledge that if their careers were threatened, it would not be owing to the outcome of a single day.

If the tone of this book seems steeped in the culture of Darwinism and evolutionary thinking, it does not come from any remotely formal training in the natural sciences, but from the evolutionary way of thinking taught by my Monte Carlo simulators.

I reckon that I outgrew the desire to generate random runs every time I want to explore an idea—but by dint of playing with a Monte Carlo engine for years I can no longer visualize a realized outcome without reference to the nonrealized ones. I call that "summing under histories," borrowing the expression from the colorful physicist Richard Feynman who applied such methods to examine the dynamics of subatomic particles.

Using my Monte Carlo to make and remake history reminded me of the experimental novels (the so-called new novels) by such writers as Alain Robbe-Grillet, popular in the 1960s and 1970s. There the same chapter would be written and revised, the writer each time changing the plot like a new sample path. Somehow the author was freed from the past situation he helped create and allowed himself the indulgence to change the plot retroactively.

Denigration of History

One more word on history seen from a Monte Carlo perspective. The wisdom of such classical stories as Solon's prods me to spend even more time in the company of the classical historians, even if the stories, like Solon's warning, have benefited from the patina of time. However, this goes against the grain: Learning from history does not come naturally to us humans, a fact that is so visible in the endless repetitions of identically configured booms and busts in modern markets. By history I refer to the anecdotes, not the historical theorizing, the grand-scale historicism that aims to interpret events with theories based on uncovering some laws in the evolution of history—the sort of Hegelianism and pseudoscientific historicism leading to such calls as the end of history (it is pseudoscientific because it draws theories from past events without allowing for the fact that such combinations of events might have arisen from randomness; there is no way to verify the claims in a controlled experiment). For me, history is of use merely at the level of my desired sensibility, affecting the way I would wish to think by reference to past events, by being able to better steal the ideas of others and leverage them, correct the mental defect that seems to block my ability to learn from others. It is the respect of the elders that I would like to develop, reinforcing the awe I instinctively feel for people with gray hair, but that has eroded in my life as a trader where age and success are somewhat divorced. Indeed, I have two ways of learning from history: from the past, by reading the elders; and from the future, thanks to my Monte Carlo toy.

The Stove Is Hot

As I mentioned above, it is not natural for us to learn from history. We have enough clues to believe that our human endowment does not favor transfers of experience in a cultural way but through selection of those who bear some favorable traits. It is a platitude that children learn only from their own mistakes; they will cease to touch a burning stove only when they are themselves burned; no possible warning by others can lead to developing the smallest form of cautiousness. Adults, too, suffer from such a condition. This point has been examined by behavioral economics pioneers Daniel Kahneman and Amos Tversky with regard to the choices people make in selecting risky medical treatments—I myself have seen it in my being extremely lax in the area of detection and prevention (i.e., I refuse to derive my risks from the probabilities computed on others, feeling that I am somewhat special) yet extremely aggressive in the treatment of medical conditions (I overreact when I am burned), which is not coherent with rational behavior under uncertainty. This congenital denigration of the experience of others is not limited to children or to people like myself; it affects business decision makers and investors on a grand scale.

If you think that merely reading history books would help you learn "from other's mistakes," consider the following nineteenth-century experiment. In a well-known psychology case the Swiss doctor Claparède had an amnesic patient completely crippled with her ailment. Her condition was so bad that he would have to reintroduce himself to her at a frequency of once per fifteen minutes for her to remember who he was. One day he secreted a pin in his hand before shaking hers. The next day she quickly withdrew her hand as he tried to greet her, *but still did not recognize him*. Since then plenty of discussions of amnesic patients show some form of learning on the part of people without their being aware of it and without it being stored in conscious memory. The scientific name of the distinction between the two memories, the conscious and the nonconscious, is declarative and nondeclarative. Much of the risk avoidance that comes from experiences is part of the second. The only way I developed a respect for history is by making myself aware of the fact that I was not programmed to learn from it in a textbook format.

Actually, things can be worse than that: In some respects we do not learn from our own history. Several branches of research have been examining our inability to learn from our own reactions to past events: For example, people fail to learn that their emotional reactions to past experiences (positive or negative) were short-lived—yet they continuously retain the bias of thinking that the purchase of an object will bring long-lasting, possibly permanent, happiness or that a setback will cause severe and prolonged distress (when in the past similar setbacks did not affect them for very long and the joy of the purchase was short-lived).

All of my colleagues who I have known to denigrate history blew up spectacularly—and I have yet to encounter some such person who has not blown up. But the truly interesting point lies in the remarkable similarities in their approaches. The blowup, I will repeat, is different from merely incurring a monetary loss; it is losing money when one does not believe that such fact is possible at all. There is nothing wrong with a risk taker taking a hit provided one declares that one is a risk taker rather than that the risk being taken is small or nonexistent. Characteristically, blown-up traders think that they knew enough about the world to reject the possibility of the adverse event taking place: There was no courage in their taking such risks, just ignorance. I have noticed plenty of analogies between those who blew up in the stock market crash of 1987, those who blew up in the Japan meltdown of 1990, those who blew up in the bond market débâcle of 1994, those who blew up in Russia in 1998, and those who blew up shorting Nasdaq stocks. They all made claims to the effect that "these times are different" or that "their market was different," and offered seemingly well-constructed, intellectual arguments (of an economic nature) to justify their claims; they were unable to accept that the experience of others was out there, in the open, freely available to all, with books detailing crashes in every bookstore. Aside from these generalized systemic blowups, I have seen hundreds of option traders forced to leave the business after blowing up in a stupid manner, in spite of warnings by the veterans, similar to a child's touching the stove. This I find to resemble my own personal attitude with respect to the detection and prevention of the variety of ailments I may be subjected to. Every man believes himself to be quite different, a matter that amplifies the "why me?" shock upon a diagnosis.

Skills in Predicting Past History

We can discuss this point from different angles. Experts call one manifestation of such denigration of history *historical determinism*. In a nutshell we think that we would know when history is made; we believe that people who, say, witnessed the stock market crash of 1929 knew then that they lived an acute historical event and that, should these events repeat themselves, they too would know about such facts. Life for us is made to resemble an adventure movie, as we know ahead of time that something big is about to happen. It is hard to imagine that people who witnessed history did not know at the time how important the moment was. Somehow all respect we may have for history does not translate well into our treatment of the present.

Jean-Patrice of the last chapter was abruptly replaced by an interesting civil servant type who had never been involved in the randomness professions. He just went to the right civil servant schools where people learn to write reports and had some senior managerial position in the institution. As is typical with subjectively assessed positions he tried to make his predecessor look bad: Jean-Patrice was deemed sloppy and unprofessional. The civil servant's first undertaking was to run a formal analysis of our transactions; he found that we traded a little too much, incurring very large back office expenditure. He analyzed a large segment of foreign exchange traders' transactions, then wrote a report explaining that only close to 1% of these transactions generated significant profits; the rest generated either losses or small profits. He was shocked that the traders did not do more of the winners and less of the losers. It was obvious to him that we needed to comply with these instructions immediately. If we just doubled the winners, the results for the institution would be so great. How come you highly paid traders did not think about it before?

Things are always obvious after the fact. The civil servant was a very intelligent person, and this mistake is much more prevalent than one would think. It has to do with the way our mind handles historical information. When you look at the past, the past will always be deterministic, since only one single observation took place. Our mind will interpret most events not with the preceding ones in mind, but the following ones. Imagine taking a test knowing the answer. While we know that history flows forward, it is difficult to realize that we envision it backward. Why is it so? We will discuss the point in Chapter 11 but here is a possible explanation: Our minds are not quite designed to understand how the world works, but, rather, to get out of trouble rapidly and have progeny. If they were made for us to understand things, then we would have a machine in it that would run the past history as in a VCR, with a correct chronology, and it would slow us down so much that we would have trouble operating. Psychologists call this overestimation of what one knew at the time of the event due to subsequent information the *hindsight bias*, the "I knew it all along" effect.

Now the civil servant called the trades that ended up as losers "gross mistakes," just like journalists call decisions that end up costing a candidate his election a "mistake." I will repeat this point until I get hoarse: A mistake is not something to be determined after the fact, but in the light of the information until that point.

A more vicious effect of such hindsight bias is that those who are very good at *predicting* the past will think of themselves as good at predicting the future, and feel confident about their ability to do so. This is why events like those of September 11, 2001, never teach us that we live in a world where important events are not predictable—even the Twin Towers' collapse appears to have been predictable *then*.

My Solon

I have another reason to be obsessed with Solon's warning. I hark back to the very same strip of land in the Eastern Mediterranean where the story took place. My ancestors experienced bouts of extreme opulence and embarrassing penury over the course of a single generation, with abrupt regressions that people around me who have the memory of steady and linear betterment do not think feasible (at least not at the time of writing). Those around me either have (so far) had few family setbacks (except for the Great Depression) or, more generally, are not suffused with enough sense of history to reflect backward. For people of my background, Eastern Mediterranean Greek Orthodox and invaded Eastern Roman citizens, it was as if our soul had been wired with the remembrance of that sad spring day circa 500 years ago when Constantinople, under the invading Turks, fell out of history, leaving us the lost subjects of a dead empire, very prosperous minorities in an Islamic world—but with an extremely fragile wealth. Moreover, I vividly remember the image of my own dignified grandfather, a former deputy prime minister and son of a deputy prime minister (whom I never saw without a suit), residing in a nondescript apartment in Athens, his estate having been blown up during the Lebanese civil war. Incidentally, having experienced the ravages of war, I find undignified impoverishment far harsher than physical danger (somehow dying in full dignity appears to me far preferable to living a janitorial life, which is one of the reasons I dislike financial risks far more than physical ones). I am certain that Croesus worried more about the loss of his Kingdom than the perils to his life.

There is an important and nontrivial aspect of historical thinking, perhaps more applicable to the markets than anything else: Unlike many "hard" sciences, history cannot lend itself to experimentation. But somehow, overall, history is potent enough to deliver, on time, in the medium to long run, most of the possible scenarios, and to eventually bury the bad guy. Bad trades catch up with you, it is frequently said in the markets. Mathematicians of probability give that a fancy name: *ergodicity*. It means, roughly, that (under certain conditions) very long sample paths would end up resembling each other. The properties of a very, very long sample path would be similar to the Monte Carlo properties of an average of shorter ones. The janitor in Chapter 1 who won the lottery, if he lived one thousand years, cannot be expected to win more lotteries. Those who were unlucky in life in spite of their skills would eventually rise. The lucky fool might have benefited from some luck in life; over the longer run he would slowly converge to the state of a less-lucky idiot. Each one would revert to his long-term properties.

DISTILLED THINKING ON YOUR PALMPILOT

Breaking News

The journalist, my *bête noire*, entered this book with George Will dealing with random outcomes. In the next step I will show how my Monte Carlo toy taught me to favor distilled thinking, by which I mean the thinking based on information around us that is stripped of meaningless but diverting clutter. For the difference between noise and information, the topic of this book (noise has more randomness) has an analog: that between journalism and history. To be competent, a journalist should view matters like a historian, and play down the value of the information he is providing, such as by saying: "Today the market went up, but this information is not too relevant as it emanates mostly from noise." He would certainly lose his job by trivializing the value of the information in his hands. Not only is it difficult for the journalist to think more like a historian, but it is, alas, the historian who is becoming more like the journalist.

For an idea, age is beauty (it is premature to discuss the mathematics of the point). The applicability of Solon's warning to a life in randomness, in contrast with the exact opposite message delivered by the prevailing media-soaked culture, reinforces my instinct to value distilled thought over newer thinking, regardless of its apparent sophistication—another reason to accumulate the hoary volumes by my bedside (I confess that the only news items I currently read are the far more interesting upscale social gossip stories found in *Tatler, Paris Match*, and *Vanity Fair*—in addition to *The Economist*). Aside from the decorum of ancient thought as opposed to the coarseness of fresh ink, I have spent some time phrasing the idea in the mathematics of evolutionary arguments and conditional probability. For an idea to have survived so long across so many cycles is indicative of its relative fitness. Noise, at least *some* noise, was filtered out. Mathematically, progress means that some new information is better than past information, not that the average of new information will supplant past information, which means that it is optimal for someone, when in doubt, to systematically reject the new idea, information, or method. Clearly and shockingly, always. Why?

The argument in favor of "new things" and even more "new new things" goes as follows: Look at the dramatic changes that have been brought about by the arrival of new technologies, such as the automobile, the airplane, the telephone, and the personal computer. Middlebrow inference (inference stripped of probabilistic thinking) would lead one to believe that all new technologies and inventions would likewise revolutionize our lives. But the answer is not so obvious: Here we only see and count the winners, to the exclusion of the losers (it is like saying that actors and writers are rich, ignoring the fact that actors are largely waiters—and lucky to be ones, for the less comely writers usually serve French fries at McDonald's). Losers? The Saturday newspaper lists dozens of new patents of such items that can revolutionize our lives. People tend to infer that because *some* inventions have revolutionized our lives that inventions are good to endorse and we should favor the new over the old. I hold the opposite view. The opportunity cost of missing a "new new thing" like the airplane and the automobile is minuscule compared to the toxicity of all the garbage one has to go through to get to these jewels (assuming these have brought some improvement to our lives, which I frequently doubt).

Now the exact same argument applies to information. The problem with information is not that it is diverting and generally useless, but that it is toxic. We will examine the dubious value of the highly frequent news with a more technical discussion of signal filtering and observation frequency farther down. I will say here that such respect for the time-honored provides arguments to rule out any commerce with the babbling modern journalist and implies a minimal exposure to the media as a guiding principle for someone involved in decision making under uncertainty. If there is anything better than noise in the mass of "urgent" news pounding us, it would be like a needle in a haystack. People do not realize that the media is paid to get your attention. For a journalist, silence rarely surpasses any word.

On the rare occasions when I boarded the 6:42 train to New York I observed with amazement the hordes of depressed business commuters (who seemed to prefer to be elsewhere) studiously buried in *The Wall Street Journal*, apprised of the minutiae of companies that, at the time of writing now, are probably out of business. Indeed it is difficult to ascertain whether they seem depressed because they are reading the newspaper, or if depressive people tend to read the newspaper, or if people who are living outside their genetic habitat both read the newspaper and look sleepy and depressed. But while early on in my career such focus on noise would have offended me intellectually, as I would have deemed such information as too statistically insignificant for the derivation of any meaningful conclusion, I currently look at it with delight. I am happy to see such mass-scale idiotic decision making, prone to overreaction in their postperusal investment orders—in other words I currently see in the fact that people read such material an insurance for my continuing in the entertaining business of option trading against the fools of randomness. (It takes a huge invest-ment in introspection to learn that the thirty or more hours spent "studying" the news last month neither had any predictive ability during your activities of that month nor did it impact your current knowledge of the world. This problem is similar to the weaknesses in our ability to correct for past errors: Like a health club mem-bership taken out to satisfy a New Year's resolution, people often think that it will surely be the next batch of news that will really make a difference to their understanding of things.)

Shiller Redux

Much of the thinking about the negative value of information on society in general was sparked by Robert Shiller. Not just in financial markets; but overall his 1981 paper may be the first mathematically formulated introspection on the manner in which society in general handles information. Shiller made his mark with his 1981 paper on the volatility of markets, where he determined that if a stock price is the estimated value of "something" (say the discounted cash flows from a corporation), then market prices are way too volatile in relation to tangible manifestations of that "something" (he used dividends as proxy). Prices swing more than the fundamentals they are supposed to reflect, they visibly overre-act by being too high at times (when their price overshoots the good news or when they go up without any marked reason) or too low at others. The volatility differential between prices and infor-mation meant that something about "rational expectation" did not work. (Prices did not rationally reflect the long-term value of se-curities and were overshooting in either direction.) Markets had to be wrong. Shiller then pronounced markets to be not as efficient as established by financial theory (efficient markets meant, in a nutshell, that prices should adapt to all available information in such a way as to be totally unpredictable to us humans and pre-vent people from deriving profits). This conclusion set off calls by

the religious orders of high finance for the destruction of the infi-
del who committed such apostasy. Interestingly, and by some
strange coincidence, it is that very same Shiller who was trounced
by George Will only one chapter ago.

The principal criticism against Shiller came from Robert C.
Merton. The attacks were purely on methodological grounds
(Shiller's analysis was extremely rough; for instance, his using dividends in place of earnings was rather weak). Merton was also defending the official financial theory position that markets needed
to be efficient and could not possibly deliver opportunities on a
silver plate. Yet the same Robert C. Merton later introduced himself as the "founding partner" of a hedge fund that aimed at taking
advantage of market inefficiencies. Setting aside the fact that Merton's hedge fund blew up rather spectacularly from the *black swan
problem* (with characteristic denial), his "founding" such a hedge
fund requires, by implication, that he agrees with Shiller about the
inefficiency of the market. The defender of the dogmas of modern
finance and efficient markets started a fund that took advantage of
market inefficiencies! It is as if the Pope converted to Islam.

Things are not getting any better these days. At the time of
writing, news providers are offering all manner of updates, "breaking news" that can be delivered electronically in a wireless manner.
The ratio of undistilled information to distilled is rising, saturating
markets. The elder's messages need not be delivered to you as im-
minent news.

This does not mean that all journalists are fooled by randomness noise providers: There are hordes of thoughtful journalists in
the business (I would suggest London's Anatole Kaletsky and New
York's Jim Grant and Alan Abelson as the underrated representa-
tives of such a class among financial journalists; Gary Stix among
scientific journalists); it is just that prominent media journalism is
a thoughtless process of providing the noise that can capture people's attention and there exists no mechanism for separating the

two. As a matter of fact, smart journalists are often penalized. Like the lawyer in Chapter 11 who does not care about the truth, but about arguments that can sway a jury whose intellectual defects he knows intimately, journalism goes to what can capture our attention, with adequate sound bites. Again, my scholarly friends would wonder why I am getting emotional stating the obvious things about the journalists; the problem with my profession is that we depend on them for what information we need to obtain.

Gerontocracy

A preference for distilled thinking implies favoring old investors and traders, that is, investors who have been exposed to markets the longest, a matter that is counter to the common Wall Street practice of preferring those that have been the most profitable, and preferring the youngest whenever possible. I toyed with Monte Carlo simulations of heterogeneous populations of traders under a variety of regimes (closely resembling historical ones), and found a significant advantage in selecting aged traders, using as a selection criterion their cumulative years of experience rather than their absolute success (conditional on their having survived without blowing up). "Survival of the fittest," a term so hackneyed in the investment media, does not seem to be properly understood: Under regime switching, as we will see in Chapter 5, it will be unclear who is actually the fittest, and those who will survive are not necessarily those who appear to be the fittest. Curiously, it will be the oldest, simply because older people have been exposed longer to the rare event and can be, convincingly, more resistant to it. I was amused to discover a similar evolutionary argument in mate selection that considers that women prefer (on balance) to mate with healthy older men over healthy younger ones, everything else being equal, as the former provide some evidence of better genes. Gray hair signals an enhanced ability to survive—conditional on having reached the gray hair stage, a man is likely to be more resistant to the vagaries of life. Curiously, life insurers in renaissance Italy reached the same conclusion, by charging the same insurance for a man in his twenties as they did for a man in his fifties, a sign that they had the same life expectancy; once a man crossed the forty-year mark, he had shown that very few ailments could harm him. We now proceed to a mathematical rephrasing of these arguments.

PHILOSTRATUS IN MONTE CARLO: ON THE DIFFERENCE BETWEEN NOISE AND INFORMATION

The wise man listens to meaning; the fool only gets the noise. The modern Greek poet C. P. Cavafy wrote a piece in 1915 after Philostratus' adage "For the gods perceive things in the future, ordinary people things in the present, but the wise perceive things about to happen." Cavafy wrote:

> In their intense meditation the hidden sound of things approaching reaches them and they listen reverently while in the street outside the people hear nothing at all.

I thought hard and long on how to explain with as little mathematics as possible the difference between noise and meaning, and how to show why the time scale is important in judging a historical event. The Monte Carlo simulator can provide us with such an intuition. We will start with an example borrowed from the investment world, as it can be explained rather easily, but the concept can be used in any application.

Let us manufacture a happily retired dentist, living in a pleasant, sunny town. We know *a priori* that he is an excellent investor, and that he will be expected to earn a return of 15% in excess of Treasury bills, with a 10% error rate per annum (what we call volatility). It means that out of 100 sample paths, we expect close to 68 of them to fall within a band of plus and minus 10% around the 15% excess return, i.e., between 5% and 25% (to be technical; the bell-shaped normal distribution has 68% of all observations falling between −1 and 1 standard deviations). It also means that 95 sample paths would fall between −5% and 35%.

Clearly, we are dealing with a very optimistic situation. The dentist builds for himself a nice trading desk in his attic, aiming to spend every business day there watching the market, while sipping decaffeinated cappuccino. He has an adventurous temperament, so he finds this activity more attractive than drilling the teeth of reluctant little old Park Avenue ladies.

He subscribes to a Web-based service that supplies him with continuous prices, now to be obtained for a fraction of what he pays for his coffee. He puts his inventory of securities in his spreadsheet and can thus instantaneously monitor the value of his speculative portfolio. We are living in the era of connectivity.

A 15% return with a 10% volatility (or uncertainty) per annum translates into a 93% probability of success in any given year. But seen at a narrow time scale, this translates into a mere 50.02% probability of success over any given second as shown in Table 3.1. Over the very narrow time increment, the observation will reveal

Table 3.1 Probability of success at different scales

Scale	Probability
1 year	93%
1 quarter	77%
1 month	67%
1 day	54%
1 hour	51.3%
1 minute	50.17%
1 second	50.02%

close to nothing. Yet the dentist's heart will not tell him that. Being emotional, he feels a pang with every loss, as it shows in red on his screen. He feels some pleasure when the performance is positive, but not in equivalent amount as the pain experienced when the performance is negative.

At the end of every day the dentist will be emotionally drained. A minute-by-minute examination of his performance means that each day (assuming eight hours per day) he will have 241 pleasurable minutes against 239 unpleasurable ones. These amount to 60,688 and 60,271, respectively, per year. Now realize that if the unpleasurable minute is worse in reverse pleasure than the pleasurable minute is in pleasure terms, then the dentist incurs a large deficit when examining his performance at a high frequency.

Consider the situation where the dentist examines his portfolio only upon receiving the monthly account from the brokerage house. As 67% of his months will be positive, he incurs only four pangs of pain per annum and eight uplifting experiences. This is the same dentist following the same strategy. Now consider the dentist looking at his performance only every year. Over the next 20 years that he is expected to live, he will experience 19 pleasant surprises for every unpleasant one!

This scaling property of randomness is generally misunderstood, even by professionals. I have seen Ph.D.s argue over a performance observed in a narrow time scale (meaningless by any standard). Before additional dumping on the journalist, more observations seem in order.

Viewing it from another angle, if we take the ratio of noise to what we call nonnoise (i.e., left column/right column), which we have the privilege here of examining quantitatively, then we have the following. Over one year we observe roughly 0.7 parts noise for every one part performance. Over one month, we observe roughly 2.32 parts noise for every one part performance. Over one hour, 30 parts noise for every one part performance, and over one second, 1,796 parts noise for every one part performance.

A few conclusions:

1. Over a short time increment, one observes the variability of the portfolio, not the returns. In other words, one sees the variance, little else. I always remind myself that what one observes is at best a combination of variance and returns, not just returns (but my emotions do not care about what I tell myself).

2. Our emotions are not designed to understand the point. The dentist did better when he dealt with monthly statements rather than more frequent ones. Perhaps it would be even better for him if he limited himself to yearly statements. (If you think that you can control your emotions, think that some people also believe that they can control their heartbeat or hair growth.)

3. When I see an investor monitoring his portfolio with live prices on his cellular telephone or his handheld, I smile and smile.

Finally, I reckon that I am not immune to such an emotional defect. But I deal with it by having no access to information, except in rare circumstances. Again, I prefer to read poetry. If an event is important enough, it will find its way to my ears. I will return to this point in time.

The same methodology can explain why the news (the high scale) is full of noise and why history (the low scale) is largely stripped of it (though fraught with interpretation problems). This explains why I prefer not to read the newspaper (outside of the obituary), why I never chitchat about markets, and, when in a trading room, I frequent the mathematicians and the secretaries, not the traders. It explains why it is better to read *The New Yorker* on Mondays than *The Wall Street Journal* every morning (from the standpoint of frequency, aside from the massive gap in intellectual class between the two publications).

Finally, this explains why people who look too closely at randomness burn out, their emotions drained by the series of pangs they experience. Regardless of what people claim, a negative pang is not offset by a positive one (some psychologists estimate the negative effect for an average loss to be up to 2.5 the magnitude of a positive one); it will lead to an emotional deficit.

Now that you know that the high-frequency dentist has more exposure to both stress and positive pangs, and that these do not cancel out, consider that people in lab coats have examined some scary properties of this type of negative pangs on the neural system (the usual expected effect: high blood pressure; the less expected: chronic stress leads to memory loss, lessening of brain plasticity, and brain damage). To my knowledge there are no studies investigating the exact properties of trader's burnout, but a daily exposure to such high degrees of randomness without much control will have physiological effects on humans (nobody studied the effect of such exposure on the risk of cancer). What economists did not understand for a long time about positive and negative kicks is that both their biology and their intensity are different. Consider that they are mediated in different parts of the brain—and that the degree of rationality in decisions made subsequent to a gain is extremely different from the one after a loss.

Note also that the implication that wealth does not count so much into one's well-being as the route one uses to get to it.

Some so-called wise and rational persons often blame me for "ignoring" possible valuable information in the daily newspaper and refusing to discount the details of the noise as "short-term events." Some of my employers have blamed me for living on a different planet.

My problem is that I am not rational and I am extremely prone to drown in randomness and to incur emotional torture. I am aware of my need to ruminate on park benches and in cafés away from information, but I can only do so if I am somewhat deprived of it. My sole advantage in life is that I know some of my weaknesses, mostly that I am incapable of taming my emotions facing news and incapable of seeing a performance with a clear head. Silence is far better. More on that in Part III.

Four

RANDOMNESS, NONSENSE, AND THE SCIENTIFIC INTELLECTUAL

On extending the Monte Carlo generator to produce artificial thinking and compare it with rigorous nonrandom constructs. The science wars enter the business world. Why the aesthete in me loves to be fooled by randomness.

RANDOMNESS AND THE VERB

Our Monte Carlo engine can take us into more literary territory. Increasingly, a distinction is being made between the scientific intellectual and the literary intellectual—culminating with what is called the "science wars," plotting factions of literate nonscientists against no less literate scientists. The distinction between the two approaches originated in Vienna in the 1930s, with a collection of physicists who decided that the large gains in science were becoming significant enough to make claims on the field known to belong to the humanities. In their view, literary thinking could conceal plenty of well-sounding nonsense. They wanted to strip thinking from rhetoric (except in literature and poetry where it properly belonged).

The way they introduced rigor into intellectual life is by declaring that a statement could fall only into two categories: *deductive*, like "2 + 2 = 4," i.e., incontrovertibly flowing from a precisely defined axiomatic framework (here the rules of arithmetic), or *inductive*, i.e., verifiable in some manner (experience, statistics, etc.), like "it rains in Spain" or "New Yorkers are generally rude." Anything else was plain unadulterated hogwash (music could be a far better replacement to metaphysics). Needless to say that inductive statements may turn out to be difficult, even impossible, to verify, as we will see with the black swan problem—and empiricism can be worse than any other form of hogwash when it gives someone confidence (it will take me a few chapters to drill the point). However, it was a good start to make intellectuals responsible for providing some form of evidence for their statements. This Vienna Circle was at the origin of the development of the ideas of Popper, Wittgenstein (in his later phase), Carnap, and flocks of others. Whatever merit their original ideas may have, the impact on both philosophy and the practice of science has been significant. Some of their impact on nonphilosophical intellectual life is starting to develop, albeit considerably more slowly.

One conceivable way to discriminate between a scientific intellectual and a literary intellectual is by considering that a scientific intellectual can usually recognize the writing of another but that the literary intellectual would not be able to tell the difference between lines jotted down by a scientist and those by a glib nonscientist. This is even more apparent when the literary intellectual starts using scientific buzzwords, like "uncertainty principle," "Gödel's theorem," "parallel universe," or "relativity," either out of context or, as often, in exact opposition to the scientific meaning. I suggest reading the hilarious *Fashionable Nonsense* by Alan Sokal for an illustration of such practice (I was laughing so loudly and so frequently while reading it on a plane that other passengers kept whispering things about me). By dumping the kitchen sink of scientific references in a paper, one can make another literary intellectual believe that one's material has the stamp of science. Clearly, to a scientist, science lies in the rigor of the inference, not in random references to such grandiose concepts as general relativity or quantum indeterminacy. Such rigor can be spelled out in plain English. Science is method and rigor; it can be identified in the simplest of prose writing. For instance, what struck me while reading Richard Dawkins' *Selfish Gene* is that, although the text does not exhibit a single equation, it seems as if it were translated from the language of mathematics. Yet it is artistic prose.

Reverse Turing Test

Randomness can be of considerable help with the matter. For there is another, far more entertaining way to make the distinction between the babbler and the thinker. You can sometimes replicate something that can be mistaken for a literary discourse with a Monte Carlo generator but it is not possible randomly to construct a scientific one. Rhetoric can be constructed randomly, but not genuine scientific knowledge. This is the application of *Turing's test* of artificial intelligence, except in reverse. What is the Turing test? The brilliant British mathematician, eccentric, and computer pioneer Alan Turing came up with the following test: A computer can be said to be intelligent if it can (on average) fool a human into mistaking it for another human. The converse should be true. A human can be said to be unintelligent if we can replicate his speech by a computer, which we know is unintelligent, and fool a human into believing that it was written by a human. Can one produce a piece of work that can be largely mistaken for Derrida entirely randomly?

The answer seems to be yes. Aside from the hoax by Alan Sokal (the same of the hilarious book a few lines ago), who managed to produce nonsense and get it published by some prominent journal, there are Monte Carlo generators designed to structure such texts and write entire papers. Fed with "postmodernist" texts, they can randomize phrases under a method called recursive grammar, and produce grammatically sound but entirely meaningless sentences that sound like Jacques Derrida, Camille Paglia, and such a crowd. Owing to the fuzziness of his thought, the literary intellectual can be fooled by randomness.

At the Monash University program in Australia featuring the Dada Engine built by Andrew C. Bulhak, I toyed with the engine and generated a few papers containing the following sentences:

However, the main theme of the works of Rushdie is not theory, as the dialectic paradigm of reality suggests, but pretheory. The premise of the neosemanticist paradigm of discourse implies that sexual identity, ironically, has significance.

Many narratives concerning the role of the writer as observer may be revealed. It could be said that if cultural narrative holds, we have to choose between the dialectic paradigm of narrative and neoconceptual Marxism. Sartre's analysis of cultural narrative holds that society, paradoxically, has objective value.

Thus, the premise of the neodialectic paradigm of expression implies that consciousness may be used to reinforce hierarchy, but only if reality is distinct from consciousness; if that is not the case, we can assume that language has intrinsic meaning.

Some business speeches belong to this category in their own right, except that they are less elegant and draw on a different type of vocabulary than the literary ones. We can randomly construct a speech imitating that of your chief executive officer to ensure whether what he is saying has value, or if it is merely dressed-up nonsense from someone who was lucky to be put there. How? You select randomly five phrases below, then connect them by adding the minimum required to construct a grammatically sound speech.

> We look after our customer's interests / the road ahead / our assets are our people / creation of shareholder value / our vision / our expertise lies in / we provide interactive solutions / we position ourselves in this market / how to serve our customers better / short-term pain for long-term gain / we will be rewarded in the long run / we play from our strength and improve our weaknesses / courage and determination will prevail / we are committed to innovation and technology / a happy employee is a productive employee / commitment to excellence / strategic plan / our work ethics.

If this bears too close a resemblance to the speech you just heard from the boss of your company, then I suggest looking for a new job.

The Father of All Pseudothinkers

It is hard to resist discussion of artificial history without a comment on the father of all pseudothinkers, Hegel. Hegel writes a jargon that is meaningless outside of a chic Left Bank Parisian café or the humanities department of some university extremely well insulated from the real world. I suggest this passage from the German "philosopher" (this passage was detected, translated, and reviled by Karl Popper):

> Sound is the change in the specific condition of segregation of the material parts, and in the negation of this condition; merely an abstract or an ideal ideality, as it were, of that specification. But this change, accordingly, is itself immediately the negation of the material specific subsistence; which is, therefore, real ideality of specific gravity and cohesion, i.e.—heat. The heating up of

sounding bodies, just as of beaten and or rubbed ones, is the appearance of heat, originating conceptually together with sound.

Even a Monte Carlo engine could not sound as random as the great philosophical master thinker (it would take plenty of sample runs to get the mixture of "heat" and "sound." People call that philosophy and frequently finance it with taxpayer subsidies! Now consider that Hegelian thinking is generally linked to a "scientific" approach to history; it has produced such results as Marxist regimes and even a branch called "neo-Hegelian" thinking. These "thinkers" should be given an undergraduate-level class on statistical sampling theory prior to their release into the open world.

MONTE CARLO POETRY

There are instances where I like to be fooled by randomness. My allergy to nonsense and verbiage dissipates when it comes to art and poetry. On the one hand, I try to define myself and behave officially as a no-nonsense hyperrealist ferreting out the role of chance; on the other, I have no qualms indulging in all manner of personal superstitions. Where do I draw the line? The answer is aesthetics. Some aesthetic forms appeal to something in our biology, whether or not they originate in random associations or plain hallucination. Something in our human genes is deeply moved by the fuzziness and ambiguity of language; then why fight it?

The poetry and language lover in me was initially depressed by the account of the "exquisite cadavers" poetic exercise, where interesting and poetic sentences are randomly constructed. By throwing enough words together, some unusual and magical-sounding metaphor is bound to emerge according to the laws of combinatorics. Yet one cannot deny that some of these poems are of ravishing beauty. Who cares about their origin if they manage to please our aesthetic senses?

The story of the "exquisite cadavers" is as follows. In the aftermath of the First World War, a collection of surrealist poets—which included André Breton, their pope, Paul Eluard, and others—got together in cafés and tried the following exercise (modern literary critics attribute the exercise to the depressed mood after the war and the need to escape reality). On a folded piece of paper, in turn, each one of them would write a predetermined part of a sentence, not knowing the others' choice. The first would pick an adjective, the second a noun, the third a verb, the fourth an adjective, and the fifth a noun. The first publicized exercise of such random (and collective) arrangement produced the following poetic sentence:

> *The exquisite cadavers shall drink the new wine.*
> *(Les cadavres exquis boiront le vin nouveau.)*

Impressive? It sounds even more poetic in the native French. Quite impressive poetry has been produced in such a manner, sometimes with the aid of a computer. But poetry has never been truly taken seriously outside of the beauty of its associations, whether they have been produced by the random ranting of one or more disorganized brains, or the more elaborate constructions of one conscious creator.

Now, regardless of whether the poetry was obtained by a Monte Carlo engine or sung by a blind man in Asia Minor, language is potent in bringing pleasure and solace. Testing its intellectual validity by translating it into simple logical arguments would rob it of a varying degree of its potency, sometimes excessively; nothing can be more bland than translated poetry. A convincing argument of the role of language is the existence of surviving holy languages, uncorrupted by the no-nonsense tests of daily use. Semitic religions, that is Judaism, Islam, and original Christianity understood the point: Keep a language away from the rationalization of daily use and avoid the corruption of the vernacular. Four decades ago, the Catholic church translated the services and liturgies from Latin to the local vernaculars; one may wonder if this caused a drop in religious beliefs. Suddenly religion subjected itself to being judged by intellectual and scientific, without the aesthetic, standards. The Greek Orthodox church made the lucky mistake, upon translating some of its prayers from Church Greek into the Semitic-based vernacular spoken by the Grecosyrians of the Antioch region (southern Turkey and northern Syria), of choosing classical Arabic, an entirely dead language. My folks are thus lucky to pray in a mixture of dead Koiné (Church Greek) and no less dead Koranic Arabic.

What does this point have to do with a book on randomness? Our human nature dictates a need for *péché mignon*. Even the economists, who usually find completely abstruse ways to escape reality, are starting to understand that what makes us tick is not necessarily the calculating accountant in us. We do not need to be rational and scientific when it comes to the details of our daily life—only in those that can harm us and threaten our survival. Modern life seems to invite us to do the exact opposite; become extremely realistic and intellectual when it comes to such matters as religion and personal behavior, yet as irrational as possible when it comes to matters ruled by randomness (say, portfolio or real estate investments). I have encountered colleagues, "rational," no-nonsense people, who do not understand why I cherish the poetry of Baudelaire and Saint-John Perse or obscure (and often impenetrable) writers like Elias Canetti, J. L. Borges, or Walter Benjamin. Yet they get sucked into listening to the "analyses" of a television "guru," or into buying the stock of a company they know absolutely nothing about, based on tips by neighbors who drive expensive cars. The Vienna Circle, in their dumping on Hegel-style verbiage-based philosophy, explained that, from a scientific standpoint, it was plain garbage, and, from an artistic point of view, it was inferior to music. I have to say that I find Baudelaire far more pleasant to frequent than CNN newscasters or listening to George Will.

There is a Yiddish saying: "If I am going to be forced to eat pork, it better be of the best kind." If I am going to be fooled by randomness, it better be of the beautiful (and harmless) kind. This point will be made again in Part III.

•

Five

SURVIVAL OF THE LEAST FIT—CAN EVOLUTION BE FOOLED BY RANDOMNESS?

A case study on two rare events. On rare events and evolution. How "Darwinism" and evolution are concepts that are misunderstood in the nonbiological world. Life is not continuous. How evolution will be fooled by randomness. A prolegomenon for the problem of induction.

CARLOS THE EMERGING-MARKETS WIZARD

I used to meet Carlos at a variety of New York parties, where he would show up impeccably dressed, though a bit shy with the ladies. I used to regularly pounce on him and try to pick his brains about what he did for a living, namely buying or selling emerging-market bonds. A nice gentleman, he complied with my requests, but tensed up; for him speaking English, in spite of his fluency, seemed to require some expenditure of physical effort that made him contract his head and neck muscles (some people are not made to speak foreign languages). What are emerging-market bonds? "Emerging market" is the politically correct euphemism to define a country that is not very developed (as a skeptic, I do not impart to their "emergence" such linguistic certainty). The bonds are financial instruments issued by these foreign governments, mostly Russia, Mexico, Brazil, Argentina, and Turkey. These bonds traded for pennies when these governments were not doing well. Suddenly investors rushed into these markets in the early 1990s and pushed the envelope further and further by acquiring increasingly more exotic securities. All these countries were building hotels where United States cable news channels were available, with health clubs equipped with treadmills and large-screen television sets that made them join the global village. They all had access to the same gurus and financial entertainers. Bankers would come to invest in their bonds and the countries would use the proceeds to build nicer hotels so more investors would visit. At some point these bonds became the vogue and went from pennies to dollars; those who knew the slightest thing about them accumulated vast fortunes.

Carlos supposedly comes from a patrician Latin-American family that was heavily impoverished by the economic troubles of the 1980s, but, again, I have rarely run into anyone from a ravaged country whose family did not at some juncture own an entire province or, say, supply the Russian czar with sets of dominoes. After brilliant undergraduate studies, he went to Harvard to pursue a Ph.D. in economics, as it was the sort of thing Latin-American patricians had gotten into the habit of doing at the time (with a view to saving their economies from the evils of non-Ph.D. hands). He was a good student but could not find a decent thesis topic for his dissertation. Nor did he gain the respect of his thesis advisor, who found him unimaginative. Carlos settled for a master's degree and a Wall Street career.

The nascent emerging-market desk of a New York bank hired Carlos in 1992. He had the right ingredients for success; he knew where on the map to find the countries that issued "Brady bonds," dollar-denominated debt instruments issued by Less Developed Countries. He knew what Gross Domestic Product meant. He looked serious, brainy, and well-spoken, in spite of his heavy Spanish accent. He was the kind of person banks felt comfortable putting in front of their customers. What a contrast with the other traders who lacked polish!

Carlos got there right in time to see things happening in that market. When he joined the bank, the market for emerging-market debt instruments was small and traders were located in undesirable parts of trading floors. But the activity rapidly became a large, and growing, part of the bank's revenues.

He was generic among this community of emerging market traders; they are a collection of cosmopolitan patricians from across the emerging-market world that remind me of the international coffee hour at the Wharton School. I find it odd that rarely does a person specialize in the market of his or her birthplace. Mexicans based in London trade Russian securities, Iranians and Greeks specialize in Brazilian bonds, and Argentines trade Turkish securities. Unlike my experience with real traders, they are generally urbane, dress well, collect art, but are nonintellectual. They seem too conformist to be true traders. They are mostly between thirty and forty, owing to the youth of their market. You can expect many of them to hold season tickets to the Metropolitan Opera. True traders, I believe, dress sloppily, are often ugly, and exhibit the intellectual curiosity of someone who would be more interested in the information-revealing contents of the garbage can than the Cézanne painting on the wall.

Carlos thrived as a trader-economist. He had a large network of friends in the various Latin-American countries and knew exactly what took place there. He bought bonds that he found attractive, either because they paid him a good rate of interest, or because he believed that they would become more in demand in the future, therefore appreciating in price. It would be perhaps erroneous to call him a *trader*. A trader buys and sells (he may sell what he does not own and buy it back later, hopefully making a profit in a decline; this is called "shorting"). Carlos just bought—and he bought in size. He believed that he was paid a good risk premium to hold these bonds because there was economic value in lending to these countries. Shorting, in his opinion, made no economic sense.

Within the bank Carlos was the emerging-markets reference. He could produce the latest economic figures at the drop of a hat. He had frequent lunches with the chairman. In his opinion, trading was economics, little else. It had worked so well for him. He got promotion after promotion, until he became the head trader of the emerging-market desk at the institution. Starting in 1995, Carlos did exponentially well in his new function, getting an expansion of his capital on a steady basis (i.e., the bank allocated a larger portion of its funds to his operation)—so fast that he was incapable of using up the new risk limits.

The Good Years

The reason Carlos had good years was not just because he bought emerging-market bonds and their value went up over the period. It was mostly because he also bought dips. He accumulated when prices experienced a momentary panic. The year 1997 would have been bad had he not added to his position after the dip in October that accompanied the false stock market crash that took place then. Overcoming these small reversals of fortune made him feel invincible. He could do no wrong. He believed that the economic intuition he was endowed with allowed him to make good trading decisions. After a market dip he would verify the fundamentals, and, if they remained sound, he would buy more of the security and lighten up as the market recovered. Looking back at the emerging-market bonds between the time Carlos started his involvement with these markets and his last bonus check in December 1997, one sees an upward sloping line, with occasional blips, such as the Mexican devaluation of 1995, followed by an extended rally. One can also see some occasional dips that turned out to be "excellent buying opportunities."

It was the summer of 1998 that undid Carlos—that last dip did not translate into a rally. His track record up to that point included just one bad quarter—but bad it was. He had earned for his bank close to $80 million cumulatively in his previous years. He lost $300 million in just one summer. What happened? When the market started dipping in June, his friendly sources informed him that the sell-off was merely the result of a "liquidation" by a New Jersey hedge fund run by a former Wharton professor. That fund specialized in mortgage securities and had just received instructions to wind down the overall inventory. The inventory included some Russian bonds, mostly because *yield hogs*, as these funds are known, engage in the activity of building a "diversified" portfolio of high-yielding securities.

Averaging Down

When the market started falling, he accumulated more Russian bonds, at an average of around $52. That was Carlos' trait, average down. The problems, he deemed, had nothing to do with Russia, and it was not some New Jersey fund run by some mad scientist that was going to decide the fate of Russia. "Read my lips: It's a li-qui-da-tion!" he yelled at those who questioned his buying.

By the end of June, his trading revenues for 1998 had dropped from up $60 million to up $20 million. That made him angry. But he calculated that should the market rise back to the pre–New Jersey sell-off, then he would be up $100 million. That was unavoidable, he asserted. These bonds, he said, would never, ever trade below $48. He was risking so little, to possibly make so much.

Then came July. The market dropped a bit more. The benchmark Russian bond was now at $43. His positions were underwater, but he increased his stakes. By now he was down $30 million for the year. His bosses were starting to become nervous, but he kept telling them that, after all, Russia would not go under. He repeated the cliché that it was too big to fail. He estimated that bailing them out would cost so little and would benefit the world economy so much that it did not make sense to liquidate his inventory now. "This is the time to buy, not to sell," he said repeatedly. "These bonds are trading very close to their possible default value." In other words, should Russia go into default, and run out of dollars to pay the interest on its debt, these bonds would hardly budge. Where did he get this idea? From discussions with other traders and emerging-market economists (or trader-economist hybrids). Carlos put about half his net worth, then $5 million, in the Russia Principal Bond. "I will retire on these profits," he told the stockbroker who executed the trade.

Lines in the Sand

The market kept going through the lines in the sand. By early August, they were trading in the thirties. By the middle of August, they were in the twenties. And he was taking no action. He felt that the price on the screen was quite irrelevant in his business of buying "value."

Signs of battle fatigue were starting to show in his behavior. Carlos was getting jumpy and losing some of his composure. He yelled at someone in a meeting: "Stop losses are for schmucks! I am not going to buy high and sell low!" During his string of successes he had learned to put down and berate traders of the non-emerging-market variety. "Had we gotten out in October 1997 after our heavy loss we would not have had those excellent 1997 results," he was also known to repeat. He also told management: "These bonds trade at very depressed levels. Those who can invest now in these markets would realize wonderful returns." Every morning, Carlos spent an hour discussing the situation with market economists around the globe. They all seemed to present a similar story: This sell-off is overdone.

Carlos' desk experienced losses in other emerging markets as well. He also lost money in the domestic Russian Ruble Bond market. His losses were mounting, but he kept telling his management rumors about very large losses among other banks—larger than his. He felt justified to show that "he fared well relative to the industry." This is a symptom of systemic troubles; it shows that there was an entire community of traders who were conducting the exact same activity. Such statements, that other traders had also gotten into trouble, are self-incriminating. A trader's mental construction should direct him to do precisely *what other people do not do.*

Toward the end of August, the bellwether Russia Principal Bonds were trading below $10. Carlos' net worth was reduced by almost half. He was dismissed. So was his boss, the head of trading. The president of the bank was demoted to a "newly created position." Board members could not understand why the bank had so much exposure to a government that was not paying its own employees—which, disturbingly, included armed soldiers. This was one of the small points that emerging-market economists around the globe, from talking to each other so much, forgot to take into account. Veteran trader Marty O'Connell calls this the firehouse effect. He had observed that firemen with much downtime who talk to each other for too long come to agree on many things that an outside, impartial observer would find ludicrous (they develop political ideas that are very similar). Psychologists give it a fancier name, but my friend Marty has no training in behavioral sciences.

The nerdy types at the International Monetary Fund had been taken for a ride by the Russian government, which cheated on its account. Let us remember that economists are evaluated on how intelligent they sound, not on a scientific measure of their knowledge of reality. However, the price of the bonds was not fooled. It knew more than the economists, more than the Carloses of the emerging-market departments.

Louie, a veteran trader on the neighboring desk who suffered much humiliation by these rich emerging-market traders, was there, vindicated. Louie was then a fifty-two-year-old Brooklyn-born-and-raised trader who over three decades survived every single conceivable market cycle. He calmly looked at Carlos being escorted by a security guard to the door like a captured soldier taken to the arena. He muttered in his Brooklyn accent: *"Economics Schmeconomics. It is all market dynamics."*

Carlos is now out of the market. The possibility that history may prove him right (at some point in the future) has nothing to do with the fact that he is a bad trader. He has all of the traits of a thoughtful gentleman, and would be an ideal son-in-law. But he has most of the attributes of the bad trader. And, at any point in time, the richest traders are often the worst traders. This, I will call the *cross-sectional problem*: At a given time in the market, the most successful traders are likely to be those that are best fit to the latest cycle. This does not happen too often with dentists or pianists—because these professions are more immune to randomness.

JOHN THE HIGH-YIELD TRADER

We met John, Nero's neighbor, in Chapter 1. At the age of thirty-five he had been on Wall Street as a corporate high-yield bonds trader for seven years, since his graduation from Pace University's Lubin School of Business. He rose to head up a team of ten traders in record time—thanks to a jump between two similar Wall Street firms that afforded him a generous profit-sharing contract. The contract allowed him to be paid 20% of his profits, as they stood at the end of each calendar year. In addition, he was allowed to invest his own personal money in his trades—a great privilege.

John is not someone who can be termed as principally intelligent, but he was believed to be endowed with a good measure of business sense. He was said to be "pragmatic" and "professional." He gave the impression that he was born a businessperson, never saying anything remotely unusual or out of place. He remained calm in most circumstances, rarely betraying any form of emotion. Even his occasional cursing (this is Wall Street!) was so much in context that it sounded, well, professional.

John dressed impeccably. This was in part due to his monthly trips to London where his unit had a satellite supervising European high-yield activities. He wore a Savile Row tailored dark business suit, with a Ferragamo tie—enough to convey the impression that he was the epitome of the successful Wall Street professional. Each time Nero ran into him he came away feeling poorly dressed.

John's desk engaged principally in an activity called "high-yield" trading, which consisted in acquiring "cheap" bonds that yielded, say, 10%, while the borrowing rate for his institution was 5.5%. It netted a 4.5% revenue, also called *interest rate differential*—which seemed small except that he could leverage himself and multiply such profit by the leverage factor. He did this in various countries, borrowing at the local rate and investing in "risky" assets. It was easy for him to amass over $3 billion dollars in face value of such trade across a variety of continents. He hedged the interest rate exposure by selling U.S., U.K., French, and other government bond futures, thus limiting his bet to the differential between the two instruments. He felt protected by this hedging strategy—cocooned (or so he thought) against those nasty fluctuations in the world's global interest rates.

The Quant Who Knew Computers and Equations

John was assisted by Henry, a foreign quant whose English was incomprehensible, but who was believed to be at least equally competent in risk-management methods. John knew no math; he relied on Henry. "His brains and my business sense," he was wont to say. Henry supplied him with risk assessments concerning the overall portfolio. Whenever John felt worried, he would ask Henry for another freshly updated report. Henry was a graduate student in Operations Research when John hired him. His specialty was a field called Computational Finance, which, as its name indicates, seems to focus solely on running computer programs overnight. Henry's income went from $50,000 to $600,000 in three years.

Most of the profit John generated for the institution was not attributable to the interest rate differential between the instruments described above. It came from the changes in the value of the securities John held, mostly because many other traders were acquiring them to imitate John's trading strategy (thus causing the price of these assets to rise). The interest rate differential was getting closer to what John believed was "fair value." John believed that the methods he used to calculate "fair value" were sound. He was backed by an entire department that helped him analyze and determine which bonds were attractive and offered capital appreciation potential. It was normal for him to be earning these large profits over time.

John made steady income for his employers, perhaps even better than steady. Every year the revenues he generated almost doubled as compared to the previous year. During his last year, his income experienced a quantum leap as he saw the capital allocated to his trades swell beyond his wildest expectations. His bonus check was for $10 million (pretax, which would generate close to a $5 million total tax bill). John's personal net worth reached $1 million at the age of thirty-two. By the age of thirty-five it had exceeded $16 million. Most of it came from the accumulation of bonuses—but a sizeable share came from profits on his personal portfolio. Of the $16 million, about $14 million he insisted in keeping invested in his business. They allowed him, thanks to the leverage (i.e., use of borrowed money), to keep a portfolio of $50 million involved in his trades, with $36 million borrowed from the bank. The effect of the leverage is that a small loss would be compounded and would wipe him out.

It took only a few days for the $14 million to turn into thin air—and for John to lose his job at the same time. As with Carlos, it all happened during the summer of 1998, with the meltdown of high-yield bond values. Markets went into a volatile phase during which nearly everything he had invested in went against him *at the same time.* His hedges no longer worked out. He was mad at Henry for not having figured out that these events could happen. Perhaps there was a bug in the program.

His reaction to the first losses was, characteristically, to ignore the market. "One would go crazy if one were to listen to the mood swings of the market," he said. What he meant by that statement was that the "noise" was mean reverting, and would likely be off-set by "noise" in the opposite direction. That was the translation in plain English of what Henry explained to him. But the "noise" kept adding up in the same direction.

As in a biblical cycle, it took seven years to make John a hero and just seven days to make him a failure. John is now a pariah; he is out of a job and his telephone calls are not returned. Many of his friends were in the same situation. How? With all that information available to him, his perfect track record (and therefore, in his eyes, an above-average intelligence and skill-set), and the benefit of sophisticated mathematics, how could he have failed? Is it perhaps possible that he forgot about the shadowy figure of randomness?

It took a long time for John to figure out what had happened, owing to the rapidity with which the events unfolded and his state of shellshock. The dip in the market was not very large. It was just that his leverage was enormous. What was more shock-ing for him was that all their calculations gave the event a proba-bility of 1 in 1,000,000,000,000,000,000,000,000,000 years. Henry called that a "ten sigma" event. The fact that Henry doubled the odds did not seem to matter. It made the probability 2 in 1,000,000,000,000,000,000,000,000,000 years.

When will John recover from the ordeal? Probably never. The

reason is not because John lost money. Losing money is something good traders are accustomed to. It is because he blew up; he lost more than he planned to lose. His personal confidence was wiped out. But there is another reason why John may never recover. The reason is that John was never skilled in the first place. He is one of those people who happened to be there when it all happened. He may have looked the part but there are plenty of people who look the part.

Following the incident, John regarded himself "ruined"; yet his net worth is still close to $1 million, which could be the envy of more than 99.9% of the inhabitants of our planet. Yet there is a difference between a wealth level reached from *above* and a wealth reached from *below*. The road from $16 million to $1 million is not as pleasant as the one from 0 to $1 million. In addition, John is full of shame; he still worries about running into old friends on the street.

His employer should perhaps be most unhappy with the overall outcome. John pulled some money out of the episode, the $1 million he had saved. He should be thankful that the episode did not cost him anything—except the emotional drain. His net worth did not become negative. That was not the case for his last employer. John had earned for the employers, New York investment banks, around $250 million in the course of the seven years. He lost more than $600 million for his last employer in barely a few days.

The Traits They Shared

The reader needs to be warned that not all of the emerging-market and high-yield traders talk and behave like Carlos and John. Only the most successful ones, alas, or perhaps those who were the most successful during the 1992–1998 bull cycle.

At their age, both John and Carlos still have the chance to make a career. It would be wise for them to look outside of their current profession. The odds are that they will not survive the incident. Why? Because by discussing the situation with each of them, one can rapidly see that they share the traits of the *acute successful randomness fool* who, in addition, operates in the most random of environments. What is more worrisome is that their bosses and employers shared the same trait. They, too, are permanently out of the market. We will see throughout this book what characterizes the trait. Again, there may not be a clear definition for it, but you can recognize it when you see it. No matter what John and Carlos do, they will remain fools of randomness.

A REVIEW OF MARKET FOOLS
OF RANDOMNESS CONSTANTS

Most of the traits partake of the same Table P.1 right column–left column confusion; how they are fooled by randomness. Below is a brief outline of them:

An overestimation of the accuracy of their beliefs in some measure, either economic (Carlos) or statistical (John). They never considered that the fact that trading on economic variables has worked in the past may have been merely coincidental, or, perhaps even worse, that economic analysis was fit to past events to mask the random element in it. Consider that of all the possible economic theories available, one can find a plausible one that explains the past, or a portion of it. Carlos entered the market at a time when it worked, but he never tested for periods when markets did the opposite of sound economic analysis. There were periods when economics failed traders, and others when it helped them.

The U.S. dollar was overpriced (i.e., the foreign currencies were undervalued) in the early 1980s. Traders who used their economic intuitions and bought foreign currencies were wiped out. But later those who did so got rich (members of the first crop were bust). It is random! Likewise, those who shorted Japanese stocks in the late 1980s suffered the same fate—few survived to recoup their losses during the collapse of the 1990s. Toward the end of the last century there was a group of operators called "macro" traders who dropped like flies, with, for instance, "legendary" (rather, lucky) investor Julian Robertson closing shop in 2000 after having been a star until then. Our discussion of survivorship bias will enlighten us further, but, clearly, there is nothing less rigorous than their seemingly rigorous use of economic analyses to trade.

A tendency to get married to positions. There is a saying that bad traders divorce their spouse sooner than abandon their positions. Loyalty to ideas is not a good thing for traders, scientists—or anyone.

The tendency to change their story. They become investors "for the long haul" when they are losing money, switching back and forth between traders and investors to fit recent reversals of fortune. The difference between a trader and an investor lies in the duration of the bet, and the corresponding size. There is absolutely nothing wrong with investing "for the long haul," provided one does not mix it with short-term trading—it is just that many people become long-term investors after they lose money, postponing their decision to sell as part of their denial.

No precise game plan ahead of time as to what to do in the event of losses. They simply were not aware of such a possibility. Both bought more bonds after the market declined sharply, but not in response to a predetermined plan.

Absence of critical thinking expressed in absence of revision of their stance with "stop losses." Middlebrow traders do not like selling when it is "even better value." They did not consider that perhaps their method of determining value is wrong, rather than the market failing to accommodate their measure of value. They may be right, but, perhaps, some allowance for the possibility of their methods being flawed was not made. For all his flaws, we will see that Soros seems rarely to examine an unfavorable outcome without testing his own framework of analysis.

Denial. When the losses occurred there was no clear acceptance of what had happened. The price on the screen lost its reality in favor of some abstract "value." In classic denial mode, the usual "this is only the result of liquidation, distress sales" was proffered. They continuously ignored the message from reality.

How could traders who made every single mistake in the book become so successful? Because of a simple principle concerning randomness. This is one manifestation of the survivorship bias. We tend to think that traders were successful *because* they are good. Perhaps we have turned the causality on its head; we consider them good just because they make money. One can make money in the financial markets totally out of randomness.

Both Carlos and John belong to the class of people who benefited from a market cycle. It was not merely because they were involved in the right markets. It was because they had a bent in their style that closely fitted the properties of the rallies experienced in their market during the episode. They were *dip buyers*. That happened, in hindsight, to be the trait that was the most desirable between 1992 and the summer of 1998 in the specific markets in which the two men specialized. Most of those who happened to have that specific trait, over the course of that segment of history, dominated the market. Their score was higher and they replaced people who, perhaps, were better traders.

NAIVE EVOLUTIONARY THEORIES

The stories of Carlos and John illustrate how bad traders have a short- and medium-term survival advantage over good traders. Next we take the argument to a higher level of generality. One must be either blind or foolish to reject the theories of Darwinian self-selection. However, the simplicity of the concept has drawn segments of amateurs (as well as a few professional scientists) into blindly believing in continuous and infallible Darwinism in all fields, which includes economics.

The biologist Jacques Monod bemoaned a couple of decades ago that everyone believes himself an expert on evolution (the same can be said about the financial markets); things have gotten worse. Many amateurs believe that plants and animals reproduce on a one-way route toward perfection. Translating the idea in social terms, they believe that companies and organizations are, thanks to competition (and the discipline of the quarterly report), irreversibly heading toward betterment. The strongest will survive; the weakest will become extinct. As to investors and traders, they believe that by letting them compete, the best will prosper and the worst will go learn a new craft (like pumping gas or, sometimes, dentistry).

Things are not as simple as that. We will ignore the basic misuse of Darwinian ideas in the fact that organizations do not reproduce like living members of nature—Darwinian ideas are about reproductive fitness, not about survival. The problem comes, as everything else in this book, from randomness. Zoologists found that once randomness is injected into a system, the results can be quite surprising: What seems to be an evolution may be merely a diversion, and possibly regression. For instance, Steven Jay Gould (who was accused of being more of a popularizer than a genuine scientist) found ample evidence of what he calls "genetic noise," or "negative mutations," thus incurring the wrath of some of his colleagues (he took the idea a little too far). An academic debate ensued, plotting Gould against colleagues like Dawkins who were considered by their peers as better trained in the mathematics of randomness. Negative mutations are traits that survive in spite of being worse, from the reproductive fitness standpoint, than the ones they replaced. However, they cannot be expected to last more than a few generations (under what is called temporal aggregation).

Furthermore, things can get even more surprising when randomness changes in shape, as with regime switches. A regime switch corresponds to situations when all of the attributes of a system change to the point of its becoming unrecognizable to the observer. Darwinian fitness applies to species developing over a very long time, not observed over a short term—time aggregation eliminates much of the effects of randomness; things (I read *noise*) balance out over the long run, as people say.

Owing to the abrupt rare events, we do not live in a world where things "converge" continuously toward betterment. Nor do things in life move *continuously* at all. The belief in continuity was ingrained in our scientific culture until the early twentieth century. It was said that *nature does not make jumps;* people quote this in well-sounding Latin: *Natura non facit saltus.* It is generally attributed to the eighteenth-century botanist Linnaeus who obviously got it all wrong. It was also used by Leibniz as a justification for calculus, as he believed that things are continuous no matter the resolution at which we look at them. Like many well-sounding "make sense" types of statements (such dynamics made perfect intellectual sense), it turned out to be entirely wrong, as it was denied by quantum mechanics. We discovered that, in the very small, particles jump (discretely) between states; they do not slide between them.

Can Evolution Be Fooled by Randomness?

We end this chapter with the following thought. Recall that some-one with only casual knowledge about the problems of random-ness would believe that an animal is at the maximum fitness for the conditions of its time. This is not what evolution means; *on average*, animals will be fit, but not every single one of them, and not at all times. Just as an animal could have survived because its sam-ple path was lucky, the "best" operators in a given business can come from a subset of operators who survived because of overfit-ness to a sample path—a sample path that was free of the evolu-tionary rare event. One vicious attribute is that the longer these animals can go without encountering the rare event, the more vul-nerable they will be to it. We said that should one extend time to infinity, then, *by ergodicity*, that event will happen with certainty—the species will be wiped out! For evolution means fitness to one and only one time series, not the average of all the possible envi-ronments.

By some viciousness of the structure of randomness, a prof-itable person like John, someone who is a pure loser in the long run and correspondingly unfit for survival, presents a high degree of eligibility in the short run and has the propensity to multiply his genes. Recall the hormonal effect on posture and its signaling effect to other potential mates. His success (or pseudosuccess owing to its fragility) will show in his features as a beacon. An in-nocent potential mate will be fooled into thinking (uncondition-ally) that he has a superior genetic makeup, until the following rare event. Solon seems to have gotten the point; but try to explain the problem to a naive business Darwinist—or your rich neighbor across the street.

•

Six

SKEWNESS AND ASYMMETRY

We introduce the concept of skewness: Why the terms "bull" and "bear" have limited meaning outside of zoology. A vicious child wrecks the structure of randomness. An introduction to the problem of epistemic opacity. The penultimate step before the problem of induction.

THE MEDIAN IS NOT THE MESSAGE

The essayist and scientist Steven Jay Gould (who, for a while, was my role model), was once diagnosed when he was in his forties with a deadly form of cancer of the lining of the stomach. The first piece of information he received about his odds of making it was that the *median* survival for the ailment is approximately eight months; information he felt akin to Isaiah's injunction to King Hezekiah to put his house in order in preparation for death.

Now, a medical diagnosis, particularly one of such severity, can motivate people to do intensive research, particularly those prolific writers like Gould who needed more time with us to complete a few book projects. The further research by Gould uncovered a very different story from the information he had initially been given; mainly that the *expected* (i.e., average) survival was considerably higher than eight months. It came to his notice that *expected* and *median* do not mean the same thing at all. Median means roughly that 50% of the people die before eight months and 50% survive longer than eight months. But those who survive would live considerably longer, generally going about life just like a regular person and fulfilling the average 73.4 or so years predicted by insurance mortality tables.

There is asymmetry. Those who die do so very early in the game, while those who live go on living very long. Whenever there is asymmetry in outcomes, the *average* survival has nothing to do with the *median* survival. This prompted Gould, who thus understood the hard way the concept of skewness, to write his heartfelt piece "The Median Is Not the Message." His point is that the concept of median used in medical research does not characterize a probability distribution.

I will simplify Gould's point by introducing the concept of *mean* (also called *average* or *expectation*) as follows, by using a less morbid example, that of gambling. I will give an example of both asymmetric odds and asymmetric outcomes to explain the point. Asymmetric odds means that probabilities are not 50% for each event, but that the probability on one side is higher than the probability on the other. Asymmetric outcomes mean that the payoffs are not equal.

Assume I engage in a gambling strategy that has 999 chances in 1,000 of making $1 (event A) and 1 chance in 1,000 of losing $10,000 (event B), as in Table 6.1. My expectation is a loss of close to $9 (obtained by multiplying the probabilities by the corresponding outcomes). The *frequency* or *probability* of the loss, in and by itself, is totally irrelevant; it needs to be judged in connection with

Table 6.1

Event	Probability	Outcome	Expectation
A	999/1000	$1	$.999
B	1/1000	–$10,000	–$10
		Total	–$9.001

the *magnitude* of the outcome. Here A is far more likely than B. Odds are that we would make money by betting for event A, but it is not a good idea to do so.

This point is rather common and simple; it is understood by anyone making a simple bet. Yet I had to struggle all my life with people in the financial markets who do not seem to internalize it. I am not talking of novices; I am talking of people with advanced degrees (albeit MBAs) who cannot come to grips with the difference.

How could people miss such a point? Why do they confuse probability and expectation, that is, probability and probability times the payoff? Mainly because much of people's schooling comes from examples in symmetric environments, like a coin toss, where such a difference does not matter. In fact, the so-called bell curve that seems to have found universal use in society is entirely symmetric. More on that later.

BULL AND BEAR ZOOLOGY

The general press floods us with concepts like *bullish* and *bearish* which refer to the effect of higher (bullish) or lower (bearish) prices in the financial markets. But also we hear people saying "I am *bullish* on Johnny" or "I am *bearish* on that guy Nassim in the back who seems incomprehensible to me," to denote the belief in the likelihood of someone's rise in life. I have to say that *bullish* or *bearish* are often hollow words with no application in a world of randomness—particularly if such a world, like ours, presents asymmetric outcomes.

When I was in the employment of the New York office of a large investment house, I was subjected on occasions to the harrying weekly "discussion meeting," which gathered most professionals of the New York trading room. I do not conceal that I was not fond of such gatherings, and not only because they cut into my gym time. While the meetings included traders, that is, people who are judged on their numerical performance, it was mostly a forum for salespeople (people capable of charming customers), and the category of entertainers called Wall Street "economists" or "strategists," who make pronouncements on the fate of the markets, but do not engage in any form of risk taking, thus having their success dependent on rhetoric rather than actually testable facts. During the discussion, people were supposed to present their opinions on the state of the world. To me, the meeting was pure intellectual pollution. Everyone had a story, a theory, and insights that they wanted others to share. I resent the person who, without having done much homework in libraries, thinks that he is onto something rather original and insightful on a given subject matter (and I respect people with scientific minds, like my friend Stan Jonas, who feel compelled to spend their nights reading wholesale on a subject matter, trying to figure out what was done on the subject by others before emitting an opinion—would the reader listen to the opinion of a doctor who does not read medical papers?).

I have to confess that my optimal strategy (to soothe my boredom and allergy to confident platitudes) was to speak as much as I could, while totally avoiding listening to other people's replies by trying to solve equations in my head. Speaking too much would help me clarify my mind, and, with a little bit of luck, I would not be "invited" back (that is, forced to attend) the following week.

I was once asked in one of those meetings to express my views on the stock market. I stated, not without a modicum of pomp, that I believed that the market would go slightly up over the next week with a high probability. How high? "About 70%." Clearly, that was a very strong opinion. But then someone interjected, "But, Nassim, you just boasted being short a very large quantity of SP500 futures, making a bet that the market would go down. What made you change your mind?" "I did not change my mind! I have a lot of faith in my bet! [Audience laughing.] As a matter of fact I now feel like selling even more!" The other employees in the room seemed utterly confused. "Are you bullish or are you bearish?" I was asked by the strategist. I replied that I could not understand the words *bullish* and *bearish* outside of their purely zoological consideration. Just as with events A and B in the preceding example, my opinion was that the market was more likely to go up ("I would be bullish"), but that it was preferable to short it ("I would be bearish"), because, in the event of its going down, it could go down a lot. Suddenly, the few traders in the room understood my opinion and started voicing similar opinions. And I was not forced to come back to the following discussion.

Let us assume that the reader shared my opinion, that the market over the next week had a 70% probability of going up and 30% probability of going down. However, let us say that it would go up by 1% on average, while it could go down by an average of 10%. What would the reader do? Is the reader *bullish* or *bearish*?

Table 6.2

Event	Probability	Outcome	Expectation
Market goes up	70%	Up 1%	0.7
Market goes down	30%	Down 10%	−3.00
		Total	−2.3

Accordingly, *bullish* or *bearish* are terms used by people who do not engage in practicing uncertainty, like the television commentators, or those who have no experience in handling risk. Alas, investors and businesses are not paid in probabilities; they are paid in dollars. Accordingly, it is not how likely an event is to happen that matters, it is how much is made when it happens that should be the consideration. How frequent the profit is irrelevant; it is the magnitude of the outcome that counts. It is a pure accounting fact that, aside from the commentators, very few people take home a check linked to how *often* they are right or wrong. What they get is a profit or loss. As to the commentators, their success is linked to how often they are right or wrong. This category includes the "chief strategists" of major investment banks the public can see on TV, who are nothing better than entertainers. They are famous, seem reasoned in their speech, plow you with numbers, but, functionally, they are there to entertain—for their predictions to have any validity they would need a statistical testing framework. Their frame is not the result of some elaborate test but rather the result of their presentation skills.

An Arrogant Twenty-nine-year-old Son

Outside of the need for entertainment in these shallow meetings I have resisted voicing a "market call" as a trader, which caused some personal strain with some of my friends and relatives. One day a friend of my father—of the rich and confident variety—called me during his New York visit (to set the elements of pecking order straight, he hinted right away during the call that he came by Concorde, with some derogatory comment on the comfort of such method of transportation). He wanted to pick my brain on the state of a collection of financial markets. I truly had no opinion, nor had made the effort to formulate any, nor was I remotely interested in markets. The gentleman kept plowing me with questions on the state of economies, on the European central banks;

these were precise questions no doubt aiming to compare my opinion to that of some other "expert" handling his account at one of the large New York investment firms. I neither concealed that I had no clue, nor did I seem sorry about it. I was not interested in markets ("yes, I am a trader") and did not make predictions, period. I went on to explain to him some of my ideas on the structure of randomness and the verifiability of market calls but he wanted a more precise statement of what the European bond markets would do by the Christmas season.

He came away under the impression that I was pulling his leg; it almost damaged the relationship between my father and his rich and confident friend. For the gentleman called him with the following grievance: "When I ask a lawyer a legal question, he answers me with courtesy and precision. When I ask a doctor a medical question, he gives me his opinion. No specialist ever gives me disrespect. Your insolent and conceited twenty-nine-year-old son is playing *prima donna* and refuses to answer me about the direction of the market!"

Rare Events

The best description of my lifelong business in the market is "skewed bets," that is, I try to benefit from rare events, events that do not tend to repeat themselves frequently, but, accordingly, present a large payoff when they occur. I try to make money infrequently, as infrequently as possible, simply because I believe that rare events are not fairly valued, and that the rarer the event, the more undervalued it will be in price. In addition to my own empiricism, I think that the counterintuitive aspect of the trade (and the fact that our emotional wiring does not accommodate it) gives me some form of advantage.

Why are these events poorly valued? Because of a psychological bias; people who surrounded me in my career were too focused on memorizing section 2 of *The Wall Street Journal* during their train ride to reflect properly on the attributes of random events. Or perhaps they watched too many gurus on television. Or perhaps they spent too much time upgrading their PalmPilot. Even some experienced trading veterans do not seem to get the point that frequencies do not matter. Jim Rogers, a "legendary" investor, made the following statement:

> I don't buy options. Buying options is another way to go to the poorhouse. Someone did a study for the SEC and discovered that 90 percent of all options expire as losses. Well, I figured out that if 90 percent of all long option positions lost money, that meant that 90 percent of all short option positions make money. If I want to use options to be bearish, I sell calls.

Visibly, the statistic that 90% of all option positions lost money is meaningless, (i.e., the *frequency*) if we do not take into account *how much* money is made on average during the remaining 10%. If we make 50 times our bet on average when the option is in the money, then I can safely make the statement that buying options is another way to go to the palazzo rather than the poorhouse. Mr. Jim Rogers seems to have gone very far in life for someone who does not distinguish between probability and expectation (strangely, he was the partner of George Soros, a complex man who thrived on rare events—more on him later).

One such rare event is the stock market crash of 1987, which made me as a trader and allowed me the luxury of becoming involved in all manner of scholarship. Nero of the smaller house in Chapter 1 aims to get out of harm's way by avoiding exposure to rare events—a mostly defensive approach. I am far more aggressive than Nero and go one step further; I have organized my career and business in such a way as to be able to benefit from them. In other words, I aim at profiting from the rare event, with my asymmetric bets.

Symmetry and Science

In most disciplines, such asymmetry does not matter. In an academic pass/fail environment, where the cumulative grade does not matter, only frequency matters. Outside of that it is the magnitude that counts. Unfortunately, the techniques used in economics are often imported from other areas—financial economics is still a young discipline (it is certainly not yet a "science"). People in most fields outside of it do not have problems eliminating extreme values from their sample, when the difference in payoff between different outcomes is not significant, which is generally the case in education and medicine. A professor who computes the average of his students' grades removes the highest and lowest observations, which he would call *outliers*, and takes the average of the remaining ones, without this being an unsound practice. A casual weather forecaster does the same with extreme temperatures—an unusual occurrence might be deemed to skew the overall result (though we will see that this may turn out to be a mistake when it comes to forecasting future properties of the ice cap). So people in finance borrow the technique and ignore infrequent events, not noticing that the effect of a rare event can bankrupt a company.

Many scientists in the physical world are also subject to such foolishness, misreading statistics. One flagrant example is in the global-warming debate. Many scientists failed to notice it in its early stages as they removed from their sample the spikes in temperature, under the belief that these were not likely to recur. It may be a good idea to take out the extremes when computing the average temperatures for vacation scheduling. But it does not work when we study the physical properties of the weather—particularly when one cares about a cumulative effect. These scientists initially ignored the fact that these spikes, although rare, had the effect of adding disproportionately to the cumulative melting of the ice cap. Just as in finance, an event, although rare, that brings large consequences cannot just be ignored.

ALMOST EVERYBODY IS ABOVE AVERAGE

Jim Rogers is not the only person committing such traditional fallacy of mistaking mean and median. In all fairness to him, some people who think for a living, such as the star philosopher Robert Nozik, have committed versions of the same mistake (Nozik, besides, was an admirable and incisive thinker; before his premature death he was perhaps the most respected American philosopher of his generation). In his book *The Nature of Rationality* he gets, as is typical with philosophers, into amateur evolutionary arguments and writes the following: "Since not more than 50 percent of the individuals can be wealthier than average." Of course, more than 50% of individuals can be wealthier than average. Consider that you have a very small number of very poor people and the rest clustering around the middle class. The mean will be lower than the median. Take a population of 10 people, 9 having a net worth of \$30,000 and 1 having a net worth of \$1,000. The average net worth is \$27,100 and 9 out of 10 people will have above average wealth.

Figure 6.1 shows a series of points starting with an initial level W_0 and ending at the period concerned W_t. It can also be seen as the performance, hypothetical or realized, of your favorite trading strategy, the track record of an investment manager, the price of a foot of average Palazzo real estate in Renaissance Florence, the price series of the Mongolian stock market, or the difference between the U.S. and Mongolian stock markets. It is composed of a given number of sequential observations W_1, W_2, etc., ordered in such a way that the one to the right comes *after* the one to the left.

If we were dealing with a deterministic world—that is, a world stripped of randomness (the right-column world in Table P.1 on page xliii), and we knew with certainty that it was the case, things

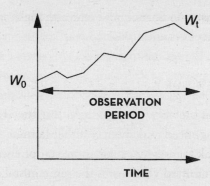

Figure 6.1 A Primer on Time Series

would be rather easy. The pattern of the series would reveal considerable and predictive information. You could tell with precision what would happen one day ahead, one year ahead, and perhaps even a decade ahead. We would not even need a statistician; a second-rate engineer would do. He does not even need to be armed with a modern degree; someone with nineteenth-century training under Laplace would be able to solve the equations, called *differential equations*, or, equivalently, *equations of motion*—since we are studying the dynamics of an entity whose position depends on time.

If we were dealing with a world where randomness is charted, things would be easy as well, given that there is an entire field created for that called *Econometrics* or *Time Series Analysis*. You would call a friendly econometrician (my experience of econometricians is that they are usually polite and friendly to practitioners). He would run the data in his software, and provide you with diagnostics that would tell you if it is worth investing in the trader generating such a track record, or if it is worth pursuing the given trading strategy. You can even buy the student version of his software for under $999 and run it yourself during the next rainy weekend.

But we are not sure that the world we live in is well charted. We will see that the judgment derived from the analysis of these past attributes may on occasion be relevant. But it may be meaningless; it could on occasion mislead you and take you in the opposite direction. Sometimes market data becomes a simple trap; it shows you the opposite of its nature, simply to get you to invest in the security or mismanage your risks. Currencies that exhibit the largest historical stability, for example, are the most prone to crashes. This was bitterly discovered in the summer of 1997 by investors who chose the safety of the pegged currencies of Malaysia, Indonesia, and Thailand (they were pegged to the U.S. dollar in a manner to exhibit no volatility, until their sharp, sudden, and brutal devaluations).

We could be either too lax or too stringent in accepting past information as a prediction of the future. As a skeptic, I reject a sole time series of the past as an indication of future performance; I need a lot more than data. My major reason is the *rare event*, but I have plenty of others.

On the surface, my statement here may seem to contradict earlier discussions, where I blame people for not learning enough from history. The problem is that we read too much into shallow recent history, with statements like "this has never happened before," but not from history in general (things that never happened before in one area tend eventually to happen). In other words, history teaches us that things that never happened before do happen. It can teach us a lot outside of the narrowly defined time series; the broader the look, the better the lesson. In other words, history teaches us to avoid the brand of naive empiricism that consists of learning from casual historical facts.

THE RARE-EVENT FALLACY

The Mother of All Deceptions

The rare event, owing to its dissimulative nature, can take a variety of shapes. It is in Mexico that it was spotted first, where it was called by academics the *peso problem*. Econometricians were puzzled by the behavior of the Mexican economic variables during the 1980s. The money supply, interest rates, or some similar measure of small relevance to the story exhibited some moody behavior, thwarting many of their efforts at modeling them. These indicators erratically switched between periods of stability and brief bursts of turbulence without warning.

By generalization, I started to label a rare event as any behavior where the adage "beware of calm waters" can hold. Popular wisdom often warns of the old neighbor who appears to remain courtly and reserved, the model of an excellent citizen, until you see his picture in the national paper as a deranged killer who went on a rampage. Until then, he was not known to have committed any transgression. There was no way to predict that such pathological behavior could emanate from such a nice person. I associate rare events with any misunderstanding of the risks derived from a narrow interpretation of past time series.

Rare events are always unexpected, otherwise they would not occur. The typical case is as follows. You invest in a hedge fund that enjoys stable returns and no volatility, until one day, you receive a letter starting with "An unforeseen and *unexpected* event, deemed a rare occurrence . . ." (emphasis mine). But rare events exist precisely because they are unexpected. They are generally caused by panics, themselves the results of liquidations (investors rushing to the door simultaneously by dumping anything they can put their hands on as fast as possible). If the fund manager or trader expected it, he and his like-minded peers would not have invested in it, and the rare event would not have taken place.

The rare event is not limited to one security. It can readily affect the performance of a portfolio. For example, many traders engage in the purchase of mortgage securities and hedge them in some manner to offset the risks and eliminate the volatility, hoping to derive some profits in excess of the Treasury bond returns (which is used as the benchmark of the minimum expected returns on an investment). They use computer programs and draw meaningful assistance from Ph.D.s in applied mathematics, astrophysics, particle physics, electrical engineering, fluid dynamics, or sometimes (though rarely) plain Ph.D.s in finance. Such a portfolio shows stable returns for long periods. Then, suddenly, as if by accident (I consider that *no* accident), the portfolio drops by 40% of its value when you expect, at the worst, a 4% drop. You call the manager to express your anger and he tells you that it was not his fault, but somehow the relationship dramatically changed (literally). He will also point out to you that similar funds also experienced the same problems.

Recall that some economists call the rare event a "peso problem." The designation peso problem does not appear to be undeservedly stereotypical. Things have not gotten better since the early 1980s with the currency of the United States' southern neighbor. Long periods of stability draw hordes of bank currency traders and hedge fund operators to the calm waters of the Mexican peso; they enjoy owning the currency because of the high interest rate it commands. Then they "unexpectedly" blow up, lose money for investors, lose their jobs, and switch careers. Then a new period of stability sets in. New currency traders come in with no memory of the bad event. They are drawn to the Mexican peso, and the story repeats itself.

It is an oddity that most fixed-income financial instruments present rare events. In the spring of 1998, I spent two hours explaining to a then-important hedge fund operator the notion of the peso problem. I went to great lengths to explain to him that the concept was generalized to every form of investment that was based on a naive interpretation of the volatility of past time series. The reply was: "You are perfectly right. We do not touch the Mexican peso. We only invest in the Russian ruble." He blew up a few months later. Until then, the Russian ruble carried attractive interest rates, which invited yield hogs of all types to get involved. He and other holders of investments denominated in rubles lost close to 97% of their investment during the summer of 1998.

We saw in Chapter 3 that the dentist does not like volatility as it causes a high incidence of negative pangs. The closer he observes his performance, the more pain he will experience owing to the greater variability at a higher resolution. Accordingly investors, merely for emotional reasons, will be drawn into strategies that experience *rare but large* variations. It is called pushing randomness under the rug. Psychologists recently found out that people tend to be sensitive to the presence or absence of a given stimulus rather than its magnitude. This implies that a loss is first perceived as just a loss, with further implications later. The same with profits. The agent would prefer the number of losses to be low and the number of gains to be high, rather than optimizing the total performance.

We can look at other aspects of the problem; think of someone involved in scientific research. Day after day, he will engage in dissecting mice in his laboratory, away from the rest of the world. He could try and try for years and years without anything to show for it. His significant other might lose patience with the *loser* who comes home every night smelling of mice urine. Until bingo, one day he comes up with a result. Someone observing the time series of his occupation would see absolutely no gain, while every day would bring him closer *in probability* to the end result.

The same with publishers; they can publish dog after dog without their business model being the least questionable, if once every decade they hit on a Harry Potter string of super-bestsellers—provided of course that they publish quality work that has a small probability of being of very high appeal. An interesting economist, Art De Vany, manages to apply these ideas to two fields: the movie business and his own health and lifestyle. He figured out the skewed properties of the movies payoffs and brought them to another level: the wild brand on nonmeasurable uncertainty we discuss in Chapter 10. What is also interesting is that he discovered that we are designed by mother nature to have an extremely skewed physical workout: Hunter-gatherers had idle moments followed by bursts of intense energy expenditure. At sixty-five, Art is said to have the physique of a man close to half his age.

In the markets, there is a category of traders who have *inverse* rare events, for whom volatility is often a bearer of good news. These traders lose money frequently, but in small amounts, and make money rarely, but in large amounts. I call them crisis hunters. I am happy to be one of them.

Why Don't Statisticians Detect Rare Events?

Statistics to the layman can appear rather complex, but the concept behind what is used today is so simple that my French mathematician friends call it deprecatorily "cuisine." It is all based on one simple notion: the more information you have, the more you are confident about the outcome. Now the problem: by how much? Common statistical method is based on the steady augmentation of the confidence level, in nonlinear proportion to the number of observations. That is, for an n times increase in the sample size, we increase our knowledge by the square root of n. Suppose I am drawing from an urn containing red and black balls. My confidence level about the relative proportion of red and black balls after 20 drawings is not twice the one I have after 10 drawings; it is merely multiplied by the square root of 2 (that is, 1.41).

Where statistics becomes complicated, and fails us, is when we have distributions that are not symmetric, like the urn above. If there is a very small probability of finding a red ball in an urn dominated by black ones, then our knowledge about the *absence* of red balls will increase very slowly—more slowly than at the expected square root of n rate. On the other hand, our knowledge of the *presence* of red balls will dramatically improve once one of them is found. This asymmetry in knowledge is not trivial; it is central in this book—it is a central philosophical problem for such people as Hume and Karl Popper (on that, later).

To assess an investor's performance, we either need more astute, and less intuitive, techniques or we may have to limit our assessments to situations where our judgment is independent of the frequency of these events.

A Mischievous Child Replaces the Black Balls

But there is even worse news. In some cases, if the incidence of red balls is itself randomly distributed, we will never get to know the composition of the urn. This is called "the problem of stationarity." Think of an urn that is hollow at the bottom. As I am sampling from it, and without my being aware of it, some mischievous child is adding balls of one color or another. My inference thus becomes insignificant. I may infer that the red balls represent 50% of the urn while the mischievous child, hearing me, would swiftly replace all the red balls with black ones. This makes much of our knowledge derived through statistics quite shaky.

The very same effect takes place in the market. We take past history as a single homogeneous sample and believe that we have considerably increased our knowledge of the future from the observation of the sample of the past. What if vicious children were changing the composition of the urn? In other words, what if things have changed?

I have studied and practiced econometrics for more than half my life (since I was nineteen), both in the classroom and in the activity of a quantitative derivatives trader. The "science" of econometrics consists of the application of statistics to samples taken at different periods of time, which we called "time series." It is based on studying the time series of economic variables, data, and other matters. In the beginning, when I knew close to nothing (that is, even less than today), I wondered whether the time series reflecting the activity of people now dead or retired should matter for predicting the future. Econometricians who knew a lot more than I did about these matters asked no such question; this hinted that it was in all likelihood a stupid inquiry. One prominent econometrician, Hashem Pesaran, answered a similar question by recommending to do "more and better econometrics." I am now convinced that, perhaps, most of econometrics could be useless— much of what financial statisticians know would not be worth knowing. For a sum of zeros, even repeated a billion times, remains zero; likewise an accumulation of research and gains in complexity will lead to naught if there is no firm ground beneath it. Studying the European markets of the 1990s will certainly be of great help to a historian; but what kind of inference can we make now that the structure of the institutions and the markets has changed so much?

Note that the economist Robert Lucas dealt a blow to econometrics by arguing that if people were rational then their rationality would cause them to figure out predictable patterns from the past and adapt, so that past information would be completely useless for predicting the future (the argument, phrased in a very mathematical form, earned him the Swedish Central Bank Prize in honor of Alfred Nobel). We are human and act according to our knowledge, which integrates past data. I can translate his point with the following analogy. If rational traders detect a pattern of stocks rising on Mondays, then, immediately such a pattern becomes detectable, it would be ironed out by people buying on Friday in anticipation of such an effect. There is no point searching for patterns that are available to everyone with a brokerage account; once detected, they would be self-canceling.

Somehow, what came to be known as the *Lucas critique* was not carried through by the "scientists." It was confidently believed that the scientific successes of the industrial revolution could be carried through into the social sciences, particularly with such movements as Marxism. Pseudoscience came with a collection of idealistic nerds who tried to create a tailor-made society, the epitome of which is the central planner. Economics was the most likely candidate for such use of science; you can disguise charlatanism under the weight of equations, and nobody can catch you since there is no such thing as a controlled experiment. Now, the spirit of such methods, called scientism by its detractors (like myself), continued past Marxism, into the discipline of finance as a few technicians thought that their mathematical knowledge could lead them to understand markets. The practice of "financial engineering" came along with massive doses of pseudoscience. Practitioners of these methods measure risks, using the tool of past history as an indication of the future. We will just say at this point that the mere possibility of the distributions not being stationary makes the entire concept seem like a costly (perhaps *very costly*) mistake. This leads us to a more fundamental question: The problem of induction, to which we will turn in the next chapter.

•

Seven

THE PROBLEM OF INDUCTION

On the chromodynamics of swans. Taking Solon's warning into some philosophical territory. How Victor Niederhoffer taught me empiricism; I added deduction. Why it is not scientific to take science seriously. Soros promotes Popper. That bookstore on Eighteenth Street and Fifth Avenue. Pascal's wager.

FROM BACON TO HUME

Now we discuss this problem viewed from the broader standpoint of the philosophy of scientific knowledge. There is a problem in inference well-known as the problem of induction. It is a problem that has been haunting science for a long time, but hard science has not been as harmed by it as the social sciences, particularly economics, even more the branch of financial economics. Why? Because the randomness content compounds its effects. Nowhere is the problem of induction more relevant than in the world of trading—and nowhere has it been as ignored!

Cygnus Atratus

In his *Treatise on Human Nature*, the Scots philosopher David Hume posed the issue in the following way (as rephrased in the now famous black swan problem by John Stuart Mill): *No amount of observations of white swans can allow the inference that all swans are white, but the observation of a single black swan is sufficient to refute that conclusion.*

Hume had been irked by the fact that science in his day (the eighteenth century) had experienced a swing from scholasticism, entirely based on deductive reasoning (no emphasis on the observation of the real world) to, owing to Francis Bacon, an overreaction into naive and unstructured empiricism. Bacon had argued against "spinning the cobweb of learning" with little practical result (science resembled theology). Science had shifted, thanks to Bacon, into an emphasis on empirical observation. The problem is that, without a proper method, empirical observations can lead you astray. Hume came to warn us against such knowledge, and to stress the need for some rigor in the gathering and interpretation of knowledge—what is called epistemology (from *episteme*, Greek for learning). Hume is the first modern *epistemologist* (epistemologists operating in the applied sciences are often called methodologists or philosophers of science). What I am writing here is not strictly true, for Hume said things far worse than that; he was an obsessive skeptic and never believed that a link between two items could be truly established as being causal. But we will tone him down a bit for this book.

Niederhoffer

The story of Victor Niederhoffer is both sad and interesting insofar as it shows the difficulty of merging extreme empiricism and logic in one single person—pure empiricism implies necessarily being fooled by randomness. I am bringing up his example because, in a way, similar to Francis Bacon, Victor Niederhoffer stood against the cobweb of learning of the University of Chicago and the efficient-market religion of the 1960s when they were at their worst. In contrast to the scholasticism of financial theorists, his work looked at data in search of anomalies and found some. He also figured out the uselessness of the news, as he showed that reading the newspaper did not confer a predictive advantage to its readers. He derived his knowledge of the world from past data stripped of preconceptions, commentaries, and stories. Since then, an entire industry of such operators, called *statistical arbitrageurs*, flourished; some of the successful ones were initially his trainees. Niederhoffer's story illustrates how empiricism cannot be inseparable from methodology.

At the center of his *modus* is Niederhoffer's dogma that any "testable" statement should be tested, as our minds make plenty of empirical mistakes when relying on vague impressions. His advice is obvious, but it is rarely practiced. How many effects we take for granted might not be there? A testable statement is one that can be broken down into quantitative components and subjected to statistical examination. For instance, a conventional-wisdom, empirical style statement like

automobile accidents happen closer to home

can be tested by taking the average distance between the accident and the domicile of the driver (if, say, about 20% of accidents happen within a twelve-mile radius). However, one needs to be careful in the interpretation; a naive reader of the result would tell you that you are more likely to have an accident if you drive in your neighborhood than if you did so in remote places, which is a typical example of naive empiricism. Why? Because accidents may happen closer to home simply because people spend their time driving close to home (if people spend 20% of their time driving in a twelve-mile radius).

But there is a more severe aspect of naive empiricism. I can use data to disprove a proposition, never to prove one. I can use history to refute a conjecture, never to affirm it. For instance, the statement

The market never goes down 20% in a given three-month period

can be tested but is completely meaningless if verified. I can quantitatively reject the proposition by finding counterexamples, but it is not possible for me to accept it simply because, in the past, the market never went down 20% in any three-month period (you cannot easily make the logical leap from "has never gone down" to "never goes down"). Samples can be greatly insufficient; markets may change; we may not know much about the market from historical information.

You can more safely use the data to reject than to confirm hypotheses. Why? Consider the following statements:

Statement A: *No swan is black, because I looked at four thousand swans and found none.*
Statement B: *Not all swans are white.*

I cannot logically make statement A, no matter how many successive white swans I may have observed in my life and may observe in the future (except, of course, if I am given the privilege of observing with certainty all available swans). It is, however, possible to make Statement B merely by finding one single counterexample. Indeed, Statement A was disproved by the discovery of Australia, as it led to the sighting of the *Cygnus atratus*, a swan variety that was jet black! The reader will see a hint of Popper's ideas, as there is a strong asymmetry between the two statements; and, furthermore, such asymmetry lies in the foundations of knowledge. It is also at the core of my operation as a decision maker under uncertainty.

I said that people rarely test testable statements; this may be better for those who cannot handle the consequence of the inference. The following inductive statement illustrates the problem of interpreting past data literally, without methodology or logic:

> I have just completed a thorough statistical examination of the life of President Bush. For fifty-eight years, close to 21,000 observations, he did not die once. I can hence pronounce him as immortal, with a high degree of statistical significance.

Niederhoffer's publicized hiccup came from his selling naked options based on his testing and assuming that what he saw in the past was an exact generalization about what could happen in the future. He relied on the statement "The market has never done this before," so he sold puts that made a small income if the statement was true and lost hugely in the event of it turning out to be wrong. When he blew up, close to a couple of decades of performance were overshadowed by a single event that only lasted a few minutes.

Another logical flaw in this type of historical statement is that often when a large event takes place, you hear the "it never happened before," as if it needed to be absent from the event's past history for it to be a surprise. So why do we consider the worst case that took place in our own past as the worst possible case? If the past, by bringing surprises, did not resemble the past previous to it (what I call the past's past), then why should our future resemble our current past?

There is another lesson to his story, perhaps the greatest one: Niederhoffer appears to approach markets as a venue from which to derive pride, status, and wins against "opponents" (such as myself), as he would in a game with defined rules. He was a squash champion with a serious competitive streak; it is just that reality does not have the same closed and symmetric laws and regulations as games. This competitive nature got him into ferocious fighting to "win." As we saw in the last chapter, markets (and life) are not simple win/lose types of situations, as the cost of the losses can be markedly different from that of the wins. Maximizing the probability of winning does not lead to maximizing the expectation from the game when one's strategy may include skewness, i.e., a small chance of large loss and a large chance of a small win. If you engaged in a Russian roulette–type strategy with a low probability of large loss, one that bankrupts you every several years, you are likely to show up as the winner in almost all samples—except in the year when you are dead.

I remind myself never to fail to acknowledge the insights of the 1960s empiricist and his early contributions. Sadly, I learned quite a bit from Niederhoffer, mostly by contrast, and particularly from the last example: not to approach anything as a *game to win*, except, of course, if it is a game. Even then, I do not like the asphyxiating structure of competitive games and the diminishing aspect of deriving pride from a numerical performance. I also learned to stay away from people of a competitive nature, as they have a tendency to commoditize and reduce the world to categories, like how many papers they publish in a given year, or how they rank in the league tables. There is something nonphilosophical about investing one's pride and ego into a "my house/library/car is bigger than that of others in my category"—it is downright foolish to claim to be first in one's category all the while sitting on a time bomb.

To conclude, extreme empiricism, competitiveness, and an absence of logical structure to one's inference can be a quite explosive combination.

SIR KARL'S PROMOTING AGENT

Next I will discuss how I discovered Karl Popper via another trader, perhaps the only one I have ever truly respected. I do not know if it applies to other people, but, in spite of my being a voracious reader, I have rarely been truly affected in my behavior (in any durable manner) by anything I have read. A book can make a strong impression, but such an impression tends to wane after some newer impression replaces it in my brain (a new book). I have to discover things by myself (recall the "Stove Is Hot" section in Chapter 3). These self-discoveries last.

One exception of ideas that stuck with me are those of Sir Karl, whom I discovered (or perhaps rediscovered) through the writings of the trader and self-styled philosopher George Soros, who seemed to have organized his life by becoming a promoter of the ideas of Karl Popper. What I learned from George Soros was not quite in the way he perhaps intended us to learn from him. I disagreed with his statements when it came to economics and philosophy. First, although I admire him greatly, I agree with professional thinkers that Soros' *forte* is not in philosophical speculation. Yet he considers himself a philosopher—which makes him endearing in more than one way. Take his first book, *The Alchemy of Finance*. On the one hand, he seems to discuss ideas of scientific explanation by throwing in big names like "deductive-nomological," something always suspicious as it is reminiscent of postmodern writers who play philosophers and scientists by using complicated references. On the other hand, he does not show much grasp of the concepts. For instance, he conducts what he calls a "trading experiment," and uses the success of the trade to imply that the theory behind it is valid. This is ludicrous: I could roll the dice to prove my religious beliefs and show the favorable outcome as evidence that my ideas are right. The fact that Soros' speculative portfolio turned a profit proves very little of anything. One cannot infer much from a single experiment in a random environment—an experiment needs a repeatability showing some causal component. Second, Soros indicts wholesale the science of economics, which may be very justified but he did not do his homework. For instance, he writes that the category of people he lumps as "economists" believe that things converge to equilibrium, when that only applies to *some* cases of neoclassical economics. There are plenty of economic theories that believe that departure from a price level can cause further divergence and cause cascading feedback loops. There has been considerable research to that effect in, say, game theory (the works of Harsanyi and Nash) or information economics (the works of Stiglitz, Akerlof, and Spence). Lumping all economics in one basket shows a bit of unfairness and lack of rigor.

But in spite of some of the nonsense in his writing, probably aimed at convincing himself that he was not just a trader, or because of it, I succumbed to the charm of this Hungarian man who like me is ashamed of being a trader and prefers his trading to be a minor extension of his intellectual life even if there is not much scholarship in his essays. Having never been impressed by people with money (and I have met plenty of those throughout my life), I did not look at any of them as remotely a role model for me. Perhaps the opposite effect holds, as I am generally repelled by the wealthy, generally because of the attitude of epic heroism that usually accompanies rapid enrichment. Soros was the only one who seemed to share my values. He wanted to be taken seriously as a Middle European professor who happened to have gotten rich owing to the validity of his ideas (it was only by failing to gain acceptance by other intellectuals that he would try to gain alpha status through his money, sort of like a seducer who, after trying hard, would end up using such an appendage as the red Ferrari to seduce the girl). In addition, although Soros did not deliver anything meaningful in his writings, he knew how to handle randomness, by keeping a critical open mind and changing his opinions with minimal shame (which carries the side effect of making him treat people like napkins). He walked around calling himself fallible, but was so potent because he knew it while others had loftier ideas about themselves. He understood Popper. Do not judge him by his writings: He lived a Popperian life.

As an aside, Popper was not new to me. I had briefly heard of Karl Popper when I was in my teens and early twenties, as part of a motivated education in Europe and the United States. But I did not understand his ideas as presented then, nor did I think it would be important (like metaphysics) for anything in life. I was at the age when one felt like one needed to read everything, which prevented one from making contemplative stops. Such hurry made it hard to detect that there was something important in Popper. It was either my conditioning by the intellectual-chic culture at the time (too much Plato, too many Marxists, too much Hegel, too many pseudoscientific intellectuals), the educational system (too many conjectures propounded as truth), or the fact that I was too young and was reading too much then to make a bridge to reality.

Popper then slipped out of my mind without hanging on a single brain cell—there was nothing in the baggage of a boy without experience to let it stick. Besides, having started trading, I entered an anti-intellectual phase; I needed to make a nonrandom buck to secure my newly lost future and wealth that had just evaporated with the Lebanese war (until then I was living with the desire to become a comfortable man of leisure like almost everyone in my family over the past two centuries). I suddenly felt financially insecure and feared becoming an employee of some firm that would turn me into a corporate slave with "work ethics" (whenever I hear *work ethics* I interpret *inefficient mediocrity*). I needed the backing of my bank account so I could buy time to think and enjoy life. The last thing I needed was immediate philosophizing and work at the local McDonald's. Philosophy, to me, became something rhetorical people did when they had plenty of time on their hands; it was an activity reserved for those who were not well versed in quantitative methods and other productive things. It was a pastime that should be limited to late hours, in bars around the campuses, when one had a few drinks and a light schedule—provided one forgot the garrulous episode as early as the next day. Too much of it can get a man in trouble, perhaps turn one into a Marxist ideologue. Popper was not to reemerge until I secured my career as a trader.

Location, Location

It is said that people generally remember the time and geographic condition where they were swept with a governing idea. The religious poet and diplomat Paul Claudel remembers the exact spot of his conversion (or reconversion) to Catholicism in the Cathedral Notre-Dame of Paris, near a precise column. Thus I remember exactly the spot at Barnes and Noble on Eighteenth Street and Fifth Avenue where in 1987, inspired by Soros, I read fifty pages of *The Open Society* and feverishly bought all the Popper titles I could get my hands on lest they run out of stock. It was in a sparsely lit side-room that had a distinctive smell of mildew. I remember vividly the thoughts that rushed through my head like a revelation.

Popper turned out to be exactly the opposite of what I initially thought about "philosophers"; he was the epitome of no nonsense. By then I had been an option trader for a couple of years and I felt angry that I was being taken for a total ride by the academic researchers in finance, particularly since I was deriving my income from the failure of their models. I had already started talking to finance academics as part of my involvement with derivatives and I had trouble getting through to them some basic points about financial markets (they believed in their models a little too much). There was all along lurking in my mind the idea that these researchers had missed a point, but I did not quite know what it was. It was not what they knew, it was how they knew it, that was the subject of my annoyance.

Popper's Answer

Popper came up with a major answer to the problem of induction (to me he came up with *the* answer). No man has influenced the way scientists do science more than Sir Karl—in spite of the fact that many of his fellow professional philosophers find him quite naive (to his credit, in my opinion). Popper's idea is that science is not to be taken as seriously as it sounds (Popper when meeting Einstein did not take him as the demigod he thought he was). There are only two types of theories:

1. Theories that are known to be wrong, as they were tested and adequately rejected (he calls them falsified).
2. Theories that have not yet been known to be wrong, not falsified yet, but are exposed to be proved wrong.

Why is a theory never *right*? Because we will never know if all the swans are white (Popper borrowed the Kantian idea of the flaws in our mechanisms of perception). The testing mechanism may be faulty. However, the statement that there is a black swan is possible to make. A theory cannot be *verified*. To paraphrase baseball coach Yogi Berra again, *past data has a lot of good in it, but it is the bad side that is bad*. It can only be provisionally accepted. A theory that falls outside of these two categories is not a theory. A theory that does not present a set of conditions under which it would be considered wrong would be termed charlatanism—it would be impossible to reject otherwise. Why? Because the astrologist can always find a reason to fit the past event, by saying that *Mars was probably in line but not too much so* (likewise to me a trader who does not have a point that would make him change his mind is not a trader). Indeed the difference between Newtonian physics, which was falsified by Einstein's relativity, and astrology lies in the following irony. Newtonian physics is scientific because it allowed us to falsify it, as we know that it is wrong, while astrology is not because it does not offer conditions under which we could reject it. Astrology cannot be disproved, owing to the auxiliary hypotheses that come into play. Such point lies at the basis of the demarcation between science and nonsense (called "the problem of demarcation").

More practically to me, Popper had many problems with statistics and statisticians. He refused to blindly accept the notion that knowledge can always increase with incremental information— which is the foundation of statistical inference. It may in some instances, but we do not know which ones. Many insightful people, such as John Maynard Keynes, independently reached the same conclusions. Sir Karl's detractors believe that favorably repeating the same experiment again and again should lead to an increased comfort with the notion that "it works." I came to understand Popper's position better once I saw the first rare event ravaging a trading room. Sir Karl feared that some type of knowledge did not increase with information—but which type we could not ascertain. The reason I feel that he is important for us traders is because to him the matter of knowledge and discovery is not so much in dealing with what we know as in dealing with what we do not know. His famous quote:

> These are men with bold ideas, but highly critical of their own ideas; they try to find whether their ideas are right by trying first to find whether they are not perhaps wrong. They work with bold conjectures and severe attempts at refuting their own conjectures.

"These" are scientists. But they could be anything.

Putting the master in context, Popper was rebelling against the growth of science. Popper intellectually came to the world with the dramatic shifts in philosophy as attempts were made to shift it from the verbal and rhetorical to the scientific and rigorous, as we saw with the presentation of the Vienna Circle in Chapter 4. These people were sometimes called the logical positivists, after the movement called *positivism* pioneered in France in the nineteenth century by Auguste Comte, where positivism meant scientification of things (literally everything under the sun). It was the equivalent of bringing the industrial revolution into the soft sciences. Without dwelling on positivism, I have to note that Popper is the antidote to positivism. To him, verification is not possible. Verificationism is more dangerous than anything else. Taken to the extreme, Popper's ideas appear naive and primitive—but they work. Note that his detractors call him a *naive falsificationist*.

I am an exceedingly naive falsificationist. Why? Because I can survive being one. My extreme and obsessive Popperism is carried out as follows. I speculate in all of my activities on theories that represent some vision of the world, but with the following stipulation: No rare event should harm me. In fact, I would like all conceivable rare events to help me. My idea of science diverges with that of the people around me walking around calling themselves scientists. Science is mere speculation, mere formulation of conjecture.

Open Society

Popper's falsificationism is intimately connected to the notion of an open society. An open society is one in which no permanent truth is held to exist; this would allow counter-ideas to emerge. Karl Popper shared ideas with his friend, the low-key economist von Hayek, who endorsed capitalism as a state in which prices can disseminate information that bureaucratic socialism would choke. Both notions of falsificationism and open society are, counterintuitively, connected to those of a rigorous method for handling randomness in my day job as a trader. Clearly, an open mind is a necessity when dealing with randomness. Popper believed that any idea of Utopia is necessarily closed owing to the fact that it chokes its own refutations. The simple notion of a good model for society that cannot be left open for falsification is totalitarian. I learned from Popper, in addition to the difference between an open and a closed society, that between an open and a closed mind.

Nobody Is Perfect

I have some sobering information about Popper the man. Witnesses of his private life find him rather un-Popperian. The philosopher and Oxford don Bryan Magee who befriended him for close to three decades depicts him as unworldly (except in his youth) and narrowly focused on his work. He spent the last fifty years of his long career (Popper lived ninety-two years) closed to the outside world, insulated from outside distractions and stimulation. Popper also engaged in giving people "firm sounding advice about their career or their private life, though he had little understanding of either. All this, of course, was in direct contravention of his professed (and indeed genuine) beliefs, and practices, in philosophy."

He was not much better in his youth. Members of the Vienna Circle tried to avoid him, not because of his divergent ideas but because he was a social problem. "He was brilliant, but self-focused, both insecure and arrogant, irascible and self-righteous. He was a terrible listener and bent on winning arguments at all costs. He had no understanding of group dynamics and no ability to negotiate them."

I will refrain from commonplace discourse about the divorce between those who have ideas and those who carry them in practice, except to bring out the interesting behavioral problem; we like to emit logical and rational ideas but we do not necessarily *enjoy* this execution. Strange as it sounds, this point has only been discovered very recently (we will see that we are not genetically fit to be rational and act rationally; we are merely fit for the maxi-

mum probability of transmitting our genes in some given unso-
phisticated environment). Also, strange as it sounds, George Soros,
obsessively self-critical, seems to be more Popperian than Popper
in his professional behavior.

Induction and Memory

Memory in humans is a large machine to make inductive inferences. Think of memories: What is easier to remember, a collection
of random facts glued together, or a story, something that offers a
series of logical links? Causality is easier to commit to memory.
Our brain would have less work to do in order to retain the information. The *size* is smaller. What is induction exactly? Induction is
going from plenty of particulars to the general. It is very handy, as
the general takes much less room in one's memory than a collection of particulars. The effect of such compression is the reduction
in the degree of detected randomness.

Pascal's Wager

I conclude with the exposition of my own method of dealing with
the problem of induction. The philosopher Pascal proclaimed that
the optimal strategy for humans is to believe in the existence of
God. For if God exists, then the believer would be rewarded. If he
does not exist, the believer would have nothing to lose. Accordingly, we need to accept the asymmetry in knowledge; there are
situations in which using statistics and econometrics can be useful.
But I do not want my life to depend on it.

Like Pascal, I will therefore state the following argument. If the
science of statistics can benefit me in anything, I will use it. If it
poses a threat, then I will not. I want to take the best of what the
past can give me without its dangers. Accordingly, I will use statistics and inductive methods to make aggressive bets, but I will not
use them to manage my risks and exposure. Surprisingly, all the
surviving traders I know seem to have done the same. They trade
on ideas based on some observation (that includes past history)

but, like the Popperian scientists, they make sure that the costs of being wrong are limited (and their probability is not derived from past data). Unlike Carlos and John, they know before getting involved in the trading strategy which events would prove their conjecture wrong and allow for it (recall that Carlos and John used past history both to make their bets and to measure their risk). They would then terminate their trade. This is called a *stop loss*, a predetermined exit point, a protection from the black swan. I find it rarely practiced.

THANK YOU, SOLON

Finally, I have to confess that upon finishing my writing of Part I, that writing about the genius of Solon's insight has carried an extreme effect on both my thinking and my private life. The composition of Part I made me even more confident in my withdrawal from the media and my distancing myself from other members of the business community, mostly other investors and traders for whom I am developing more and more contempt. I believe that I cannot have power over myself as I have an ingrained desire to integrate among people and cultures and would end up resembling them; by withdrawing myself entirely I can have a better control of my fate. I am currently enjoying a thrill of the classics I have not felt since childhood. I am now thinking of the next step: to recreate a low-information, more deterministic ancient time, say in the nineteenth century, all the while benefiting from some of the technical gains (such as the Monte Carlo engine), all of the medical breakthroughs, and all the gains of social justice of our age. I would then have the best of everything. This is called evolution.

•

Part II

MONKEYS ON TYPEWRITERS

Survivorship and Other Biases

f one puts an infinite number of monkeys in front of (strongly built) typewriters, and lets them clap away, there is a certainty that one of them would come out with an exact version of the *Iliad*. Upon examination, this may be less interesting a concept than it appears at first: Such probability is ridiculously low. But let us carry the reasoning one step beyond. Now that we have found that hero among monkeys, would any reader invest his life's savings on a bet that the monkey would write the *Odyssey* next?

In this thought experiment, it is the second step that is interesting. How much can past performance (here the typing of the *Iliad*) be relevant in forecasting future performance? The same applies to any decision based on past performance, merely relying on the attributes of the past time series. Think about the monkey showing up at your door with his impressive past performance. Hey, he wrote the *Iliad*.

The major problem with inference in general is that those whose profession is to derive conclusions from data often fall into the trap faster and more confidently than others. The more data we have, the more likely we are to drown in it. For common wisdom among people with a budding knowledge of probability laws is to base their decision making on the following principle: It is very unlikely for someone to perform considerably well in a consistent fashion without his doing something right. Track records therefore become preeminent. They call on the rule of the likelihood of such a successful run and tell themselves that if someone performed better than the rest in the past then there is a great chance of his performing better than the crowd in the future—and a very great one at that. But, as usual, beware the middlebrow: A small knowledge of probability can lead to worse results than no knowledge at all.

IT DEPENDS ON THE NUMBER OF MONKEYS

I do not deny that if someone performed better than the crowd in the past, there is a presumption of his ability to do better in the future. But the presumption might be weak, very weak, to the point of being useless in decision making. Why? Because it all depends on two factors: The randomness content of his profession and the number of monkeys in operation.

The initial sample size matters greatly. If there are five monkeys in the game, I would be rather impressed with the *Iliad* writer, to the point of suspecting him to be a reincarnation of the ancient poet. If there are a billion to the power one billion monkeys I would be less impressed—as a matter of fact I would be surprised if one of them did not get some well-known (but unspecified) piece of work, just by luck (perhaps Casanova's *Memoirs of My Life*). One monkey would even be expected to provide us with former vice president Al Gore's *Earth in the Balance*, perhaps stripped of the platitudes.

This problem enters the business world more viciously than other walks of life, owing to the high dependence on randomness (we have already belabored the contrast between randomness-dependent business with dentistry). The greater the number of businessmen, the greater the likelihood of one of them performing in a stellar manner just by luck. I have rarely seen anyone count the monkeys. In the same vein, few count the investors in the market in order to calculate, instead of the probability of success, the conditional probability of successful runs given the number of investors in operation over a given market history.

VICIOUS REAL LIFE

There are other aspects to the monkeys problem; in real life the other monkeys are not countable, let alone visible. They are hidden away, as one sees only the winners—it is natural for those who failed to vanish completely. Accordingly, one sees the survivors, and only the survivors, which imparts such a mistaken perception of the odds. We do not respond to probability, but to society's assessment of it. As we saw with Nero Tulip, even people with training in probability respond unintelligently to social pressure.

THIS SECTION

Part I described situations where people do not understand the rare event, and do not seem to accept either the possibility of its occurrence or the dire consequences of such occurrence. It also set out my own ideas, those that do not seem to have been explored in the literature. But a book on randomness is not complete without a presentation of what possible biases one might have aside from the deformations caused by the rare event. The business of Part II is more pedestrian; I will rapidly provide a synthesis of the biases of randomness as discussed in the now abundant literature on the subject.

These biases can be outlined as follows: (a) The survivorship biases (a.k.a. monkeys on a typewriter) arising from the fact that we see only winners and get a distorted view of the odds (Chapters 8 and 9, "Too Many Millionaires" and "Fry an Egg"), (b) the fact that luck is most frequently the reason for extreme success (Chapter 10, "Loser Takes All"), and (c) the biological handicap of our inability to understand probability (Chapter 11, "Randomness and Our Brain").

•

Eight

TOO MANY MILLIONAIRES NEXT DOOR

Three illustrations of the survivorship bias. Why very few people should live on Park Avenue. The millionaire next door has very flimsy clothes. An overcrowding of experts.

HOW TO STOP THE STING OF FAILURE

Somewhat Happy

Marc lives on Park Avenue in New York City with his wife, Janet, and their three children. He makes $500,000 a year, give or take a boom or a recession—he does not believe that the recent spurt in prosperity is here to last and has not mentally adjusted yet to his recent abrupt rise in income. A rotund man in his late forties, with spongy features that make him look ten years older than his age, he leads the seemingly comfortable (but heckled) life of a New York City lawyer. But he is on the quiet side of Manhattan residents. Marc is clearly not the man one would expect to go bar-hopping or attend late-night Tribeca and

SoHo parties. He and his wife have a country house and a rose garden and tend to be concerned, like many people of their age, mentality, and condition, with (in the following order) material comfort, health, and status. Weekdays, he does not come home until at least 9:30 p.m. and, at times, he can be found in the office at close to midnight. By the end of the week, Marc is so fatigued that he falls asleep during their three-hour drive to "the house"; and Marc spends most of Saturday lying in bed recovering and healing.

Marc grew up in a small town in the Midwest, the son of a quiet tax accountant who worked with sharp yellow pencils. His obsession with sharpness was so strong that he carried a sharpener in his pocket at all times. Marc exhibited very early signs of intelligence. He did extremely well in high school. He attended Harvard College, then Yale Law School. Not bad, one would say. Later his career took him to corporate law, where he started working on large cases for a prestigious New York law firm, with barely enough hours left for him to brush his teeth. This is not too much an exaggeration, for he ate almost all of his dinners in the office, accumulating body fat and Brownie points toward his partnership. He later became a partner within the usual seven years, but not without the usual human costs. His first wife (whom he met in college) left him, as she was tired of an absentee lawyer husband and weary of the deterioration in his conversation—but, ironically, she ended up moving in with and later marrying another New York lawyer, probably with a no-less-flat conversation, but who made her happier.

Too Much Work

Marc's body became progressively flabbier, and his bespoke suits needed periodic visits to the tailor, in spite of his occasional crash diets. After he got over the depression of the abandonment, he started dating Janet, his paralegal, and promptly married her. They had three children in quick succession, bought the Park Avenue apartment, and the country house.

Janet's immediate acquaintance is composed of the other parents of the Manhattan private school attended by their children, and their neighbors at the co-operative apartment building where they live. From a materialistic standpoint, they come at the low end of such a set, perhaps even at the exact bottom. They would be the poorest of these circles, as their co-op is inhabited by extremely successful corporate executives, Wall Street traders, and high-flying entrepreneurs. Their children's private school harbors the second set of children of corporate raiders, from their trophy wives—perhaps even the third set, if one takes into account the age discrepancy and the model-like features of the other mothers. By comparison, Marc's wife, Janet, like him, presents a homely country-home-with-a-rose-garden type of appearance.

You're a Failure

Marc's strategy of staying in Manhattan may be rational, as his demanding work hours would make it impossible for him to commute. But the costs on his wife, Janet, are monstrous. Why? Because of their relative nonsuccess—as geographically defined by their Park Avenue neighborhood. Every month or so, Janet has a crisis, giving in to the strains and humiliations of being snubbed by some other mother at the school where she picks up the children, or another woman with larger diamonds by the elevator of the co-op where they live in the smallest type of apartments (the G line). Why isn't her husband so successful? Isn't he smart and hardworking? Didn't he get close to 1600 on the SAT? Why is this Ronald Something, whose wife never even nods to Janet, worth hundreds of millions, when her husband went to Harvard and Yale and has such a high IQ and has hardly any substantial savings?

We will not get too involved in the Chekhovian dilemmas in the private lives of Marc and Janet, but their case provides a very common illustration of the emotional effect of *survivorship bias.* Janet feels that her husband is a failure, by comparison, but she is miscomputing the probabilities in a gross manner—she is using the wrong distribution to derive a rank. As compared to the general U.S. population, Marc has done very well, better than 99.5% of his compatriots. As compared to his high school friends, he did extremely well, a fact that he could have verified had he had time to attend the periodic reunions, and he would come at the top. As compared to the other people at Harvard, he did better than 90% of them (financially, of course). As compared to his law school comrades at Yale, he did better than 60% of them. But as compared to his co-op neighbors, he is at the bottom! Why? Because he chose to live among the people who have been successful, in an area that excludes failure. In other words, those who have failed do not show up in the sample, thus making him look as if he were not doing well at all. By living on Park Avenue, one does not have exposure to the losers, one only sees the winners. As we are cut to live in very small communities, it is difficult to assess our situation outside of the narrowly defined geographic confines of our habitat. In the case of Marc and Janet, this leads to considerable emotional distress; here we have a woman who married an extremely successful man but all she can see is comparative failure, for she cannot emotionally compare him to a sample that would do him justice.

Aside from the misperception of one's performance, there is a social treadmill effect: You get rich, move to rich neighborhoods, then become poor again. To that add the psychological treadmill effect; you get used to wealth and revert to a set point of satisfaction. This problem of some people never really getting to feel satisfied by wealth (beyond a given point) has been the subject of technical discussions on happiness.

Someone would rationally say to Janet: "Go read this book *Fooled by Randomness* by one mathematical trader on the deformations of chance in life; it would give you a statistical sense of perspective and would accordingly make you feel better." As an author, I would like to offer a panacea for $14.95, but I would rather say that in my best hopes it may provide an hour or so of solace. Janet may need something more drastic for relief. I have repeated that becoming more rational, or not feeling emotions of social slights, is not part of the human race, at least not with our current biology. There is no solace to be found from reasoning—as a trader I have learned something about these unfruitful efforts to reason against the grain. I would advise Janet to move out, and go live in some blue-collar neighborhood where they would feel less humiliated by their neighbors and rise in the pecking order beyond their probability of success. They could use the deformation in the opposite direction. If Janet cares about status, then I would even recommend some of these large housing blocks.

DOUBLE SURVIVORSHIP BIASES

More Experts

I recently read a bestseller called *The Millionaire Next Door*, an extremely misleading (but almost enjoyable) book by two "experts," in which the authors try to infer some attributes that are common to rich people. They examined a collection of currently wealthy people and found out that these are unlikely to lead lavish lives. They call such people the accumulators; persons ready to postpone consumption in order to amass funds. Most of the appeal of the book comes from the simple but counterintuitive fact that these are less likely to look like very rich people—it clearly costs money to look and behave rich, not to count the time demands of spending money. Leading prosperous lives is time-consuming— shopping for trendy clothes, becoming conversant in Bordeaux wines, getting to know the expensive restaurants. All these activities can put high demands on one's time and divert the subject from what should be the real preoccupation, namely the accumulation of nominal (and paper) wealth. The moral of the book is that the wealthiest are to be found among those less suspected to be wealthy. On the other hand, those who act and look wealthy subject their net worth to such a drain that they inflict considerable and irreversible damage to their brokerage account.

I will set aside the point that I see no special *heroism* in accumulating money, particularly if, in addition, the person is foolish enough to not even try to derive any tangible benefit from the wealth (aside from the pleasure of regularly counting the beans). I have no large desire to sacrifice much of my personal habits, intellectual pleasures, and personal standards in order to become a billionaire like Warren Buffett, and I certainly do not see the point of becoming one if I were to adopt Spartan (even miserly) habits and live in my starter house. Something about the praise lavished upon him for living in austerity while being so rich escapes me; if austerity is the end, he should become a monk or a social worker—we should remember that becoming rich is a purely selfish act, not a social one. The virtue of capitalism is that society can take advantage of people's greed rather than their benevolence, but there is no need to, in addition, extol such greed as a moral (or intellectual) accomplishment (the reader can easily see that, aside from very few exceptions like George Soros, I am not impressed by people with money). Becoming rich is not directly a moral achievement, but that is not where the severe flaw in the book lies.

As we saw, the heroes of *The Millionaire Next Door* are the accumulators, people who defer spending in order to invest. It is undeniable that such strategy might work; money spent bears no fruit (except the enjoyment of the spender). But the benefits promised in the book seem grossly overstated. A finer read of their thesis reveals that their sample includes a double dose of survivorship bias. In other words, it has two compounding flaws.

Visibility Winners

The first bias comes from the fact that the rich people selected for their sample are among the lucky monkeys on typewriters. The authors made no attempt to correct their statistics with the fact that they saw only the winners. They make no mention of the "accumulators" who have accumulated the wrong things (members of my family are experts on that; those who accumulated managed to accumulate currencies about to be devalued and stocks of companies that later went bust). Nowhere do we see a mention of the fact that some people were lucky enough to have invested in the winners; these people no doubt would make their way into the book. There is a way to take care of the bias: Lower the wealth of your average millionaire by, say, 50%, on the grounds that the bias causes the average net worth of the observed millionaire to be higher by such amount (it consists in adding the effect of the losers into the pot). It would certainly modify the conclusion.

It's a Bull Market

As to the second, more serious flaw, I have already discussed the problem of induction. The story focuses on an unusual episode in history; buying its thesis implies accepting that the current returns in asset values are permanent (the sort of belief that prevailed before the great crash that started in 1929). Remember that asset prices have (still at the time of writing) witnessed the greatest bull market in history and that values did compound astronomically during the past two decades. A dollar invested in the average stock would have grown almost twenty-fold since 1982—and that is the average stock. The sample might include people who invested in stocks performing better than average. Virtually all of the subjects became rich from asset price inflation, in other words from the recent inflation in financial paper and assets that started in 1982. An investor who engaged in the same strategy during less august days for the market would certainly have a different story to tell. Imagine the book being written in 1982, after the prolonged erosion of the inflation-adjusted value of the stocks, or in 1935, after the loss of interest in the stock market.

Or consider that the United States stock market is not the only investment vehicle. Consider the fate of those who, in place of spending their money buying expensive toys and paying for ski trips, bought Lebanese lira denominated Treasury bills (as my grandfather did), or junk bonds from Michael Milken (as many of my colleagues in the 1980s did). Go back in history and imagine the accumulator buying Russian Imperial bonds bearing the signature of Czar Nicholas II and trying to accumulate further by cashing them from the Soviet government, or Argentine real estate in the 1930s (as my great-grandfather did).

The mistake of ignoring the survivorship bias is chronic, even (or perhaps especially) among professionals. How? Because we are trained to take advantage of the information that is lying in front of our eyes, ignoring the information that we do not see. At the time of writing, pension funds and insurance companies in the United States and in Europe somehow bought the argument that "in the long term equities *always* pay off 9%" and back it up with statistics. The statistics are right, but they are past history. My argument is that I can find you a security somewhere among the 40,000 available that went up twice that amount every year without fail. Should we put the social security money into it?

A brief summing up at this point: I showed how we tend to mistake one realization among all possible random histories as the most representative one, forgetting that there may be others. In a nutshell, the survivorship bias implies that *the highest performing realization will be the most visible*. Why? Because the losers do not show up.

A GURU'S OPINION

The fund management industry is populated with gurus. Clearly, the field is randomness-laden and the guru is going to fall into a trap, particularly if he has no proper training in inference. At the time of writing, there is one such guru who developed the very unfortunate habit of writing books on the subject. Along with one of his peers, he computed the success of a "Robin Hood" policy of investing with the least successful manager in a given population of managers. It consists in switching down by taking money away from the winner and allocating it to the loser. This goes against the prevailing wisdom of investing with a winning manager and taking away money from a losing one. Doing so, their "paper strategy" (i.e., as in a Monopoly game, not executed in real life) derived considerably higher returns than if they stuck to the winning manager. Their hypothetical example seemed to them to prove that one should not stay with the best manager, as we would be inclined to do, but rather switch to the worst manager, or at least such seems to be the point they were attempting to convey.

Their analysis presents one severe hitch that any graduate student should be able to pinpoint at the first reading. Their sample only had *survivors*. They simply forgot to take into account the managers who went out of business. Such a sample includes managers that were operating during the simulation, and *are still operating today*. True, their sample included managers who did poorly, but only those managers who did poorly and recovered, without getting out of business. So it would be obvious that investing with those who fared poorly at some point but recovered (with the benefit of hindsight) would yield a positive return! Had they continued to fare poorly, they would be out of business and would not be included in the sample.

How should one conduct the proper simulation? By taking a population of managers in existence, say, five years ago and running the simulation up to today. Clearly, the attributes of those who leave the population are biased toward failure; few successful people in such a lucrative business call it quits because of extreme success. Before we turn to a more technical presentation of these issues, one mention of the much idealized buzzword of optimism. Optimism, it is said, is predictive of success. Predictive? It can also be predictive of failure. Optimistic people certainly take more risks as they are overconfident about the odds; those who win show up among the rich and famous, others fail and disappear from the analyses. Sadly.

•

Nine

IT IS EASIER TO BUY AND SELL THAN FRY AN EGG

Some technical extensions of the survivorship bias. On the distribution of "coincidences" in life. It is preferable to be lucky than competent (but you can be caught). The birthday paradox. More charlatans (and more journalists). How the researcher with work ethics can find just about anything in data. On dogs not barking.

This afternoon I have an appointment with my dentist (it will mostly consist in the dentist picking my brain on Brazilian bonds). I can state with a certain level of comfort that he knows something about teeth, particularly if I enter his office with a toothache and exit it with some form of relief. It will be difficult for someone who knows literally nothing about teeth to provide me with such relief, except if he is particularly lucky on that day— or has been very lucky in his life to become a dentist while not knowing anything about teeth. Looking at his diploma on the wall, I determine that the odds that he repeatedly gave correct answers to the exam questions and performed satisfactorily on a few thousand cavities before his graduation—out of plain random-ness—are remarkably small.

Later in the evening, I go to Carnegie Hall. I can say little about the pianist; I even forgot her unfamiliar foreign-sounding name. All I know is that she studied in some Muscovite conservatory. But I can expect to get some music out of the piano. It will be rare to have someone who performed brilliantly enough in the past to get to Carnegie Hall and now turns out to have benefited from luck alone. The expectation of having a fraud who will bang on the piano, producing only cacophonous sounds, is indeed low enough for me to rule it out completely.

I was in London last Saturday. Saturdays in London are magical; bustling but without the mechanical industry of a weekday or the sad resignation of a Sunday. Without a wristwatch or a plan I found myself in front of my favorite carvings by Canova at the Victoria and Albert Museum. My professional bent immediately made me question whether randomness played a large role in the carving of these marble statues. The bodies were realistic repro-ductions of human figures, except that they were more harmo-nious and finely balanced than anything I have seen mother nature produce on its own (Ovid's *materiam superabat opus* comes to mind). Could such finesse be a product of luck?

I can practically make the same statement about anyone oper-ating in the physical world, or in a business in which the degree of randomness is low. But there is a problem in anything related to the business world. I am bothered because tomorrow, unfortu-nately, I have an appointment with a fund manager seeking my help, and that of my friends, in finding investors. He has what he claims is a *good track record*. All I can infer is that he has learned to buy and sell. And it is harder to fry an egg than buy and sell. Well . . . the fact that he made money in the past may have some relevance, but not terribly so. This is not to say that it is always the case; there are some instances in which one can trust a track record, but, alas, there are not too many of these. As the reader now knows, the fund manager can expect to be heckled by me during the presentation, particularly if he does not exhibit the minimum of humility and self-doubt that I would expect from someone practicing randomness. I will probably bombard him with questions that he may not be prepared to answer, blinded by his past results. I will probably lecture him that Machiavelli ascribed to luck at least a 50% role in life (the rest was cunning and bravura), and that was before the creation of modern markets.

In this chapter, I discuss some well-known counterintuitive properties of performance records and historical time series. The concept presented here is well-known for some of its variations under the names *survivorship bias, data mining, data snooping, over-fitting, regression to the mean*, etc., basically situations where the performance is exaggerated by the observer, owing to a misperception of the importance of randomness. Clearly, this concept has rather unsettling implications. It extends to more general situations where randomness may play a share, such as the choice of a medical treatment or the interpretation of coincidental events.

When I am tempted to suggest a possible future contribution of financial research to science in general, I adduce the analysis of data mining and the study of survivorship biases. These have been refined in finance but can extend to all areas of scientific investigation. Why is finance so rich a field? Because it is one of the rare areas of investigation where we have plenty of information (in the form of abundant price series), but no ability to conduct experiments as in, say, physics. This dependence on past data brings about its salient defects.

FOOLED BY NUMBERS

Placebo Investors

I have often been faced with questions of the sort: "Who do you think you are to tell me that I might have been plain lucky in my life?" Well, nobody really believes that he or she was lucky. My approach is that, with our Monte Carlo engine, we can manufacture purely random situations. We can do the exact opposite of conventional methods; in place of analyzing real people hunting for attributes we can create artificial ones with precisely known attributes. Thus we can manufacture situations that depend on pure, unadulterated luck, without the shadow of skills or whatever we have called nonluck in Table P.1. In other words, we can man-make pure nobodies to laugh at; they will be *by design* stripped of any shadow of ability (exactly like a placebo drug).

We saw in Chapter 5 how people may survive owing to traits that momentarily fit the given structure of randomness. Here we take a far simpler situation where *we know the structure of randomness;* the first such exercise is a finessing of the old popular saying that *even a broken clock is right twice a day.* We will take it a bit further to show that statistics is a knife that cuts on both sides. Let us use the Monte Carlo generator introduced earlier and construct a population of 10,000 fictional investment managers (the generator is not terribly necessary since we can use a coin, or even do plain algebra, but it is considerably more illustrative—and fun). Assume that they each have a perfectly fair game; each one has a 50% probability of making $10,000 at the end of the year, and a 50% probability of losing $10,000. Let us introduce an additional restriction; once a manager has a single bad year, he is thrown out of the sample, good-bye and have a nice life. Thus we will operate like the legendary speculator George Soros who was said to tell his managers gathered in a room: "Half of you guys will be out by next year" (with an Eastern European accent). Like Soros, we have extremely high standards; we are looking only for managers with an unblemished record. We have no patience for low performers.

The Monte Carlo generator will toss a coin; *heads* and the manager will make $10,000 over the year, *tails* and he will lose $10,000. We run it for the first year. At the end of the year, we expect 5,000 managers to be up $10,000 each, and 5,000 to be down $10,000. Now we run the game a second year. Again, we can expect 2,500 managers to be up two years in a row; another year, 1,250; a fourth one, 625; a fifth, 313. We have now, simply in a fair game, 313 managers who made money for five years in a row. Out of pure luck.

Meanwhile if we throw one of these successful traders into the real world we would get very interesting and helpful comments on his remarkable style, his incisive mind, and the influences that helped him achieve such success. Some analysts may attribute his achievement to precise elements among his childhood experiences. His biographer will dwell on the wonderful role models provided by his parents; we would be supplied with black-and-white pictures in the middle of the book of a great mind in the making. And the following year, should he stop outperforming (recall that his odds of having a good year have stayed at 50%) they would start laying blame, finding fault with the relaxation in his work ethics, or his dissipated lifestyle. They will find something he did before when he was successful that he has subsequently stopped doing, and attribute his failure to that. The truth will be, however, that he simply ran out of luck.

Nobody Has to Be Competent

Let's push the argument further to make it more interesting. We create a cohort that is composed exclusively of incompetent managers. We will define an incompetent manager as someone who has a negative *expected return*, the equivalent of the odds being stacked against him. We instruct the Monte Carlo generator now to draw from an urn. The urn has 100 balls, 45 black and 55 red. By drawing with replacement, the ratio of red to black balls will remain the same. If we draw a black ball, the manager will earn $10,000. If we draw a red ball, he will lose $10,000. The manager is thus expected to earn $10,000 with 45% probability, and lose

$10,000 with 55%. On average, the manager will lose $1,000 each round—but only *on average*.

At the end of the first year, we still expect to have 4,500 managers turning a profit (45% of them), the second, 45% of that number, 2,025. The third, 911; the fourth, 410; the fifth, 184. Let us give the surviving managers names and dress them in business suits. True, they represent less than 2% of the original cohort. But they will get attention. Nobody will mention the other 98%. What can we conclude?

The first counterintuitive point is that a population entirely composed of bad managers will produce a small amount of great track records. As a matter of fact, assuming the manager shows up unsolicited at your door, it will be practically impossible to figure out whether he is good or bad. The results would not markedly change even if the population were composed entirely of managers who are expected in the long run to lose money. Why? Because owing to volatility, some of them will make money. We can see here that volatility actually helps bad investment decisions.

The second counterintuitive point is that the *expectation of the maximum* of track records, with which we are concerned, depends more on the size of the initial sample than on the individual odds per manager. In other words, the number of managers with great track records in a given market depends far more on the number of people who started in the investment business (in place of going to dental school), rather than on their ability to produce profits. It also depends on the volatility. Why do I use the notion of expectation of the maximum? Because I am not concerned at all with the average track record. I will get to see only the *best* of the managers, not all of the managers. This means that we would see more "excellent managers" in 2006 than in 1998, provided the cohort of beginners was greater in 2001 than it was in 1993—I can safely say that it was.

Regression to the Mean

The "hot hand in basketball" is another example of misperception of random sequences: It is very likely in a large sample of players for one of them to have an inordinately lengthy lucky streak. As a matter of fact it is very unlikely that an unspecified player somewhere does not have an inordinately lengthy lucky streak. This is a manifestation of the mechanism called regression to the mean. I can explain it as follows:

Generate a long series of coin flips producing heads and tails with 50% odds each and fill up sheets of paper. If the series is long enough you may get eight heads or eight tails in a row, perhaps even ten of each. Yet you know that in spite of these wins the conditional odds of getting a head or a tail is still 50%. Imagine these heads and tails as monetary bets filling up the coffers of an individual. The deviation from the norm as seen in excess heads or excess tails is here entirely attributable to luck, in other words, to variance, not to the skills of the hypothetical player (since there is an even probability of getting either).

A result is that in real life, the larger the deviation from the norm, the larger the probability of it coming from luck rather than skills: Consider that even if one has 55% probability of heads, the odds of ten wins is still very small. This can be easily verified in stories of very prominent people in trading rapidly reverting to obscurity, like the heroes I used to watch in trading rooms. This applies to height of individuals or the size of dogs. In the latter case, consider that two average-sized parents produce a large litter. The largest dogs, if they diverge too much from the average, will tend to produce offspring of smaller size than themselves, and vice versa. This "reversion" for the large outliers is what has been observed in history and explained as regression to the mean. Note that the larger the deviation, the more important its effect.

Again, one word of warning: All deviations do not come from this effect, but a disproportionately large proportion of them do.

Ergodicity

To get more technical, I have to say that people believe that they can figure out the properties of the distribution from the sample they are witnessing. When it comes to matters that depend on the maximum, it is altogether another distribution that is being inferred, that of the best performers. We call the difference between the average of such distribution and the unconditional distribution of winners and losers the survivorship bias—here the fact that about 3% of the initial cohort discussed earlier will make money five years in a row. In addition, this example illustrates the properties of ergodicity, namely, that time will eliminate the annoying effects of randomness. Looking forward, in spite of the fact that these managers were profitable in the past five years, we expect them to break even in any future time period. They will fare no better than those of the initial cohort who failed earlier in the exercise. Ah, the long term.

A few years ago, when I told one A., a then Master-of-the-Universe type, that track records were less relevant than he thought, he found the remark so offensive that he violently flung his cigarette lighter in my direction. The episode taught me a lot. Remember that nobody accepts randomness in his own success, only his failure. His ego was pumped up as he was heading up a department of "great traders" who were then temporarily making a fortune in the markets and attributing the idea to the soundness of their business, their insights, or their intelligence. They subsequently blew up during the harsh New York winter of 1994 (it was the bond market crash that followed the surprise interest rate hike by Alan Greenspan). The interesting part is that several years later I can hardly find any of them still trading (ergodicity).

Recall that the survivorship bias depends on the size of the initial population. The information that a person derived some profits in the past, just by itself, is neither meaningful nor relevant. We need to know the size of the population from which he came. In other words, without knowing how many managers out there have tried and failed, we will not be able to assess the validity of the track record. If the initial population includes ten managers, then I would give the performer half my savings without a blink. If the initial population is composed of 10,000 managers, I would ignore the results. The latter situation is generally the case; these days so many people have been drawn to the financial markets. Many college graduates are trading as a first career, failing, then going to dental school.

If, as in a fairy tale, these fictional managers materialized into real human beings, one of these could be the person I am meeting tomorrow at 11:45 a.m. Why did I select 11:45 a.m.? Because I will question him about his trading style. I need to know how he trades. I will then be able to claim that I have to rush to a lunch appointment if the manager puts too much emphasis on his track record.

LIFE IS COINCIDENTAL

Next, we look at the extensions to real life of our bias in the understanding of the distribution of coincidences.

The Mysterious Letter

You get an anonymous letter on January 2 informing you that the market will go up during the month. It proves to be true, but you disregard it owing to the well-known January effect (stocks have gone up historically during January). Then you receive another one on February 1 telling you that the market will go down. Again, it proves to be true. Then you get another letter on March 1— same story. By July you are intrigued by the prescience of the anonymous person and you are asked to invest in a special off-shore fund. You pour all your savings into it. Two months later, your money is gone. You go spill your tears on your neighbor's shoulder and he tells you that he remembers that he received two such mysterious letters. But the mailings stopped at the second letter. He recalls that the first one was correct in its prediction, the other incorrect.

What happened? The trick is as follows. The con operator pulls 10,000 names out of a phone book. He mails a bullish letter to one half of the sample, and a bearish one to the other half. The following month he selects the names of the persons to whom he mailed the letter whose prediction turned out to be right, that is, 5,000 names. The next month he does the same with the remaining 2,500 names, until the list narrows down to 500 people. Of these there will be 200 victims. An investment in a few thousand dollars' worth of postage stamps will turn into several million.

An Interrupted Tennis Game

It is not uncommon for someone watching a tennis game on television to be bombarded by advertisements for funds that did (until that minute) outperform others by some percentage over some period. But, again, why would anybody advertise if he didn't happen to outperform the market? There is a high probability of the investment coming to you if its success is caused entirely by randomness. This phenomenon is what economists and insurance people call adverse selection. Judging an investment that comes to you requires more stringent standards than judging an investment you seek, owing to such selection bias. For example, by going to a cohort composed of 10,000 managers, I have 2/100 chances of finding a spurious survivor. By staying home and answering my doorbell, the chance of the soliciting party being a spurious survivor is closer to 100%.

Reverse Survivors

We have so far discussed the spurious survivor—the same logic applies to the skilled person who has the odds markedly stacked in her favor, but who still ends up going to the cemetery. This effect is the exact opposite of the survivorship bias. Consider that all one needs is two bad years in the investment industry to terminate a risk-taking career and that, even with great odds in one's favor, such an outcome is very possible. What do people do to survive? They maximize their odds of staying in the game by taking black-swan risks (like John and Carlos)—those that fare well most of the time, but incur a risk of blowing up.

The Birthday Paradox

The most intuitive way to describe the data mining problem to a nonstatistician is through what is called the birthday paradox, though it is not really a paradox, simply a perceptional oddity. If you meet someone randomly, there is a one in 365.25 chance of your sharing their birthday, and a considerably smaller one of having the exact birthday of the same year. So, sharing the same birthday would be a coincidental event that you would discuss at the dinner table. Now let us look at a situation where there are 23 people in a room. What is the chance of there being 2 people with the same birthday? About 50%. For we are not specifying which people need to share a birthday; any pair works.

It's a Small World!

A similar misconception of probabilities arises from the random encounters one may have with relatives or friends in highly unexpected places. "It's a small world!" is often uttered with surprise. But these are not improbable occurrences—the world is much larger than we think. It is just that we are not truly testing for the odds of having an encounter with one specific person, in a specific location at a specific time. Rather, we are simply testing for any encounter, with any person we have ever met in the past, and in any place we will visit during the period concerned. The probability of the latter is considerably higher, perhaps several thousand times the magnitude of the former.

When the statistician looks at the data *to test a given relationship*, say, to ferret out the correlation between the occurrence of a given event, like a political announcement, and stock market volatility, odds are that the results can be taken seriously. But when one throws the computer at data, looking for just about *any* relationship, it is certain that a spurious connection will emerge, such as the fate of the stock market being linked to the length of women's skirts. And just like the birthday coincidences, it will amaze people.

Data Mining, Statistics, and Charlatanism

What is your probability of winning the New Jersey lottery twice? One in 17 trillion. Yet it happened to Evelyn Adams, whom the reader might guess should feel particularly chosen by destiny. Using the method we developed above, researchers Percy Diaconis and Frederick Mosteller estimated at 30 to 1 the probability that someone, somewhere, in a totally unspecified way, gets so lucky!

Some people carry their data mining activities into theology— after all, ancient Mediterraneans used to read potent messages in the entrails of birds. An interesting extension of data mining into biblical exegesis is provided in *The Bible Code* by Michael Drosnin. Drosnin, a former journalist (seemingly innocent of any training in statistics), aided by the works of a "mathematician," helped "predict" the former Israeli Prime Minister Yitzhak Rabin's assassination by deciphering a bible code. He informed Rabin, who obviously did not take it too seriously. *The Bible Code* finds statistical irregularities in the Bible; these help predict some such events.

Needless to say that the book sold well enough to warrant a sequel predicting with hindsight even more such events.

The same mechanism is behind the formation of conspiracy theories. Like *The Bible Code* they can seem perfect in their logic and can cause otherwise intelligent people to fall for them. I can create a conspiracy theory by downloading hundreds of paintings from an artist or group of artists and finding a constant among all those paintings (among the hundreds of thousand of traits). I would then concoct a conspiratorial theory around a secret message shared by these paintings. This is seemingly what the author of the bestselling *The Da Vinci Code* did.

The Best Book I Have Ever Read!

My favorite time is spent in bookstores, where I aimlessly move from book to book in an attempt to make a decision as to whether to invest the time in reading it. My buying is frequently made on impulse, based on superficial but suggestive clues. Frequently, I have nothing but a book jacket as appendage to my decision making. Jackets often contain praise by someone, famous or not, or excerpts from a book review. Good praise by a famous and respected person or a well-known magazine would sway me into buying the book.

What is the problem? I tend to confuse a book review, which is supposed to be an assessment of the quality of the book, with the *best* book reviews, marred with the same survivorship biases. I mistake the distribution of the maximum of a variable with that of the variable itself. The publisher will never put on the jacket of the book anything but the best praise. Some authors go even a step beyond, taking a tepid or even unfavorable book review and selecting words in it that appear to praise the book. One such example came from one Paul Wilmott (an English financial mathematician of rare brilliance and irreverence) who managed to announce that I gave him his "first bad review," yet used excerpts from it as praise on the book jacket (we later became friends, which allowed me to extract an endorsement from him for this book).

The first time I was fooled by this bias was upon buying, when I was sixteen, *Manhattan Transfer*, a book by John Dos Passos, the American writer, based on praise on the jacket by the French writer and "philosopher" Jean-Paul Sartre, who claimed something to the effect that Dos Passos was the greatest writer of our time. This simple remark, possibly blurted out in a state of intoxication or extreme enthusiasm, caused Dos Passos to become required reading in European intellectual circles, as Sartre's remark was mistaken for a consensus estimate of the quality of Dos Passos rather than what it was, the best remark. (In spite of such interest in his work, Dos Passos has reverted to obscurity.)

The Backtester

A programmer helped me build a *backtester*. It is a software program connected to a database of historical prices, which allows me to check the hypothetical past performance of any trading rule of average complexity. I can just apply a mechanical trading rule, like buy NASDAQ stocks if they close more than 1.83% above their average of the previous week, and immediately get an idea of its past performance. The screen will flash my hypothetical track record associated with the trading rule. If I do not like the results, I can change the percentage to, say, 1.2%. I can also make the rule more complex. I will keep trying until I find something that works well.

What am I doing? The exact same task of looking for the survivor within the set of rules that can possibly work. I am *fitting* the rule on the data. This activity is called *data snooping*. The more I try, the more I am likely, by mere luck, to find a rule that worked on past data. A random series will always present some detectable pattern. I am convinced that there exists a tradable security in the Western world that would be 100% correlated with the changes in temperature in Ulan Bator, Mongolia.

To get technical, there are even worse extensions. An outstanding recent paper by Sullivan, Timmerman, and White goes further and considers that the rules that may be in use successfully today may be the result of a survivorship bias.

Suppose that, over time, investors have experimented with technical trading rules drawn from a very wide universe—in principle thousands of parameterizations of a variety of types of rules. As time progresses, the rules that happen to perform well historically receive more attention and are considered "serious contenders" by the investment community, while unsuccessful trading rules are more likely to be forgotten. . . . If enough trading rules are considered over time, some rules are bound by pure luck, even in a very large sample, to produce superior performance even if they do not genuinely possess predictive power over asset returns. Of course, inference based solely on the subset of surviving trading rules may be misleading in this context since it does not account for the full set of initial trading rules, most of which are unlikely to have underperformed.

I have to decry some excesses in backtesting that I have closely witnessed in my private career. There is an excellent product designed just for that, called Omega TradeStation, that is currently on the market, in use by tens of thousands of traders. It even offers its own computer language. Beset with insomnia, the computerized day traders become night testers plowing the data for some of its properties. By dint of throwing their monkeys on typewriters, without specifying what book they want their monkey to write, they will hit upon hypothetical gold somewhere. Many of them blindly believe in it.

One of my colleagues, a man with prestigious degrees, grew to believe in such a virtual world to the point of losing all sense of reality. Whether the modicum of common sense left in him might have rapidly vanished under the mounds of simulations, or whether he might have had none to engage in such pursuit, I cannot tell. By closely watching him I learned that what natural skepticism he may have had vanished under the weight of data—for he was extremely skeptical, but in the wrong area. Ah, Hume!

A More Unsettling Extension

Historically, medicine has operated by trial and error—in other words, statistically. We know by now that there can be entirely fortuitous connections between symptoms and treatment, and that some medications succeed in medical trials for mere random reasons. I cannot claim expertise in medicine, but have been a steady reader of a segment of the medical literature over the past half decade, long enough to be concerned with the standards, as we will see in the next chapter. Medical researchers are rarely statisticians; statisticians are rarely medical researchers. Many medical researchers are not even remotely aware of this data mining bias. True, it may only play a small role, but it is certainly present. One recent medical study links cigarette smoking to a *reduction* in breast cancer, thus conflicting with all previous studies. Logic would indicate that the result may be suspicious, the result of mere coincidence.

The Earnings Season: Fooled by the Results

Wall Street analysts, in general, are trained to find the accounting tricks that companies use to hide their earnings. They tend to (occasionally) beat the companies at that game. But they are neither trained to reflect nor to deal with randomness (nor to understand the limitations of their methods by introspecting—stock analysts have both a worse record and higher idea of their past performance than weather forecasters). When a company shows an increase in earnings once, it draws no immediate attention. Twice, and the name starts showing up on computer screens. Three times, and the company will merit some buy recommendation.

Just as with the track record problem, consider a cohort of 10,000 companies that are assumed on average to barely return the risk-free rate (i.e., Treasury bonds). They engage in all forms of volatile business. At the end of the first year, we will have 5,000 "star" companies showing an increase in profits (assuming no inflation), and 5,000 "dogs." After three years, we will have 1,250 "stars." The stock review committee at the investment house will give your broker their names as "strong buys." He will leave a voice message that he has a hot recommendation that necessitates immediate action. You will be e-mailed a long list of names. You will buy one or two of them. Meanwhile, the manager in charge of your 401(k) retirement plan will be acquiring the entire list.

We can apply the reasoning to the selection of investment categories—as if they were the managers in the example above. Assume you are standing in 1900 with hundreds of investments to look at. There are the stock markets of Argentina, Imperial Russia, the United Kingdom, Unified Germany, and plenty of others to consider. A rational person would have bought not just the emerging country of the United States, but those of Russia and Argentina as well. The rest of the story is well-known; while many of the stock markets like those of the United Kingdom and the United States fared extremely well, the investor in Imperial Russia would have no better than medium-quality wallpaper in his hands. The countries that fared well are not a large segment of the initial cohort; randomness would be expected to allow a few investment classes to fare extremely well. I wonder if those "experts" who make foolish (and self-serving) statements like "markets will always go up in any twenty-year period" are aware of this problem.

COMPARATIVE LUCK

A far more acute problem relates to the outperformance, or the comparison, between two or more persons or entities. While we are certainly fooled by randomness when it comes to a single times series, the foolishness is compounded when it comes to the comparison between, say, two people, or a person and a benchmark. Why? Because *both* are random. Let us do the following simple thought experiment. Take two individuals, say, a person and his brother-in-law, launched through life. Assume equal odds for each of good and bad luck. Outcomes: lucky-lucky (no difference between them), unlucky-unlucky (again, no difference), lucky-unlucky (a large difference between them), unlucky-lucky (again, a large difference).

I recently attended for the first time a conference of investment managers where I sat listening to a very dull presenter comparing traders. His profession is to select fund managers and package them together for investors, something called "funds of funds" and I was listening to him as he was pouring out numbers on the screen. The first revelation was that I suddenly recognized the speaker, a former colleague biologically transformed by the passage of time. He used to be crisp, energetic, and nice; he became dull, portly, and inordinately comfortable with success. (He was not rich when I knew him—can people react to money in different ways? Do some take themselves seriously while others do not?) The second revelation was that while I suspected that he was fooled by randomness, the extent had to be far greater than one could imagine, particularly with the survivorship bias. A back of the envelope calculation showed that at least 97% of what he was discussing was just noise. The fact that he was *comparing* performances made the matter far worse.

Cancer Cures

When I return home from an Asian or European trip, my jet lag often causes me to rise at a very early hour. Occasionally, though very rarely, I switch on the TV set searching for market information. What strikes me in these morning explorations is the abundance of claims by the alternative medicine vendors of the curing power of their products. These no doubt are caused by the lower advertising rates at that time. To prove their claim, they present the convincing testimonial of someone who was cured thanks to their methods. For instance, I once saw a former throat cancer patient explaining how he was saved by a combination of vitamins for sale for the exceptionally low price of $14.95—in all likelihood he was sincere (although of course compensated for his account, perhaps with a lifetime supply of such medicine). In spite of our advances, people still believe in the existence of links between disease and cure based on such information, and there is no scientific evidence that can convince them more potently than a sincere and emotional testimonial. Such testimonial does not always come from the regular guy; statements by Nobel Prize winners (in the wrong discipline) could easily suffice. Linus Pauling, a Nobel Prize winner in chemistry, was said to believe in vitamin C's medicinal properties, himself ingesting massive daily doses. With his bully pulpit, he contributed to the common belief in vitamin C's curative properties. Many medical studies, unable to replicate Pauling's claims, fell on deaf ears as it was difficult to undo the testimonial by a "Nobel Prize winner," even if he was not qualified to discuss matters related to medicine.

Many of these claims have been harmless outside of the financial profits for these charlatans—but many cancer patients may have replaced the more scientifically investigated therapies, in favor of these methods, and died as a result of their neglecting more orthodox cures (again, the nonscientific methods are gathered under what is called "alternative medicine," that is, unproven therapies, and the medical community has difficulties convincing the press that there is only one medicine and that alternative medicine is not medicine). The reader might wonder about my claims that the user of these products could be sincere, without it meaning that he was cured by the illusory treatment. The reason is something called "spontaneous remission," in which a very small minority of cancer patients, for reasons that remain entirely spec-ulative, wipe out cancer cells and recover "miraculously." Some switch causes the patient's immune system to eradicate all cancer cells from the body. These people would have been equally cured by drinking a glass of Vermont spring water or chewing on dried beef as they were by taking these beautifully wrapped pills. Fi-nally, these spontaneous remissions might not be so spontaneous; they might, at the bottom, have a cause that we are not yet so-phisticated enough to detect.

The late astronomer Carl Sagan, a devoted promoter of scien-tific thinking and an obsessive enemy of nonscience, examined the cures from cancer that resulted from a visit to Lourdes in France, where people were healed by simple contact with the holy waters, and found out the interesting fact that, of the total cancer patients who visited the place, the cure rate was, if anything, lower than the statistical one for spontaneous remissions. It was lower than the average for those who did not go to Lourdes! Should a statis-tician infer here that cancer patients' odds of surviving deterio-rates after a visit to Lourdes?

Professor Pearson Goes to Monte Carlo (Literally): Randomness Does Not Look Random!

At the beginning of the twentieth century, as we were starting to develop techniques to deal with the notion of random outcomes, several methods were designed to detect anomalies. Professor Karl Pearson (father of Egon Pearson of Neyman-Pearson fame, famil-iar to every person who sat in a statistics 101 class) devised the first test of nonrandomness (it was in reality a test of deviation from normality, which, for all intents and purposes, was the same thing). He examined millions of runs of what was called a Monte Carlo (the old name for a roulette wheel) during the month of July 1902. He discovered that, with a high degree of statistical sig-nificance (with an error of less than one to a billion), the runs were not purely random. What! The roulette wheel was not random! Professor Pearson was greatly surprised at the discovery. But this result in itself tells us nothing; we know that there is no such thing as a pure random draw, for the outcome of the draw depends on the quality of the equipment. With enough minutiae one would be able to uncover the nonrandomness somewhere (e.g., the wheel itself may not have been perfectly balanced or perhaps the spinning ball was not completely spherical). Philosophers of statistics call this the *reference case problem* to explain that there is no true attainable randomness in practice, only in theory. Besides, a manager would question whether such nonrandomness can lead to any meaningful, profitable rules. If I need to gamble $1 on 10,000 runs and expect to make $1 for my efforts, then I would do much better in the part-time employment of a janitorial agency.

But the result bears another suspicious element. Of more practical relevance here is the following severe problem about nonrandomness. Even the fathers of statistical science forgot that a random series of runs need not exhibit a pattern to look random; as a matter of fact, data that is perfectly patternless would be extremely suspicious and appear to be man-made. A single random run is bound to exhibit some pattern—if one looks hard enough. Note that Professor Pearson was among the first scholars who were interested in creating artificial random data generators, tables one could use as inputs for various scientific and engineering simulations (the precursors of our Monte Carlo simulator). The problem is that they did not want these tables to exhibit any form of regularity. Yet real randomness does not look random!

I would further illustrate the point with the study of a phenomenon well-known as cancer clusters. Consider a square with 16 random darts hitting it with equal probability of being at any place in the square. If we divide the square into 16 smaller squares, it is expected that each smaller square will contain one dart on average—but only on average. There is a very small probability of having exactly 16 darts in 16 different squares. The average grid will have more than one dart in a few squares, and no dart at all in many squares. It will be an exceptionally rare incident that no (cancer) cluster would show on the grid. Now, transpose our grid with the darts in it to overlay a map of any region. Some newspaper will declare that one of the areas (the one with more than the average of darts) harbors radiation that causes cancer, prompting lawyers to start soliciting the patients.

The Dog That Did Not Bark: On Biases in Scientific Knowledge

By the same argument, science is marred by a pernicious survivorship bias, affecting the way research gets published. In a way that is similar to journalism, research that yields no result does not make it to print. That may seem sensible, as newspapers do not have to have a screaming headline saying that nothing new is taking place (though the Bible was smart enough to declare *ein chadash tachat hashemesh*—"nothing new under the sun," providing the information that things just do recur). The problem is that a finding of absence and an absence of findings get mixed together. There may be great information in the fact that *nothing took place*. As Sherlock Holmes noted in the *Silver Blaze* case—the curious thing was that the dog did not bark. More problematic, there are plenty of scientific results that are left out of publications because they are not statistically significant, but nevertheless provide information.

I HAVE NO CONCLUSION

I am frequently asked the question "When is it truly not luck?" There are professions in randomness for which performance is low in luck, like casinos, which manage to tame randomness. In finance? Perhaps. All traders are not speculative traders: There exists a segment called market makers whose job is to derive, like bookmakers, or even like store owners, an income against a transaction. If they speculate, their dependence on the risks of such speculation remains too small compared to their overall volume. They buy at a price and sell to the public at a more favorable one, performing large numbers of transactions. Such income provides them some insulation from randomness. Such category includes floor traders on the exchanges, bank traders who "trade against order flow," money changers in the souks of the Levant. The skills involved are sometimes rare to find: Fast thinking, alertness, a high level of energy, an ability to guess from the voice of the seller her level of nervousness; those who have them make a long career (that is, perhaps a decade). They never make it big, as their income is constrained by the number of customers, but they do well probabilistically. They are, in a way, the dentists of the profession.

Outside of this very specialized bookmaker-style profession, to be honest, I am unable to answer the question of who's lucky or unlucky. I can tell that person A seems less lucky than person B, but the confidence in such knowledge can be so weak as to be meaningless. I prefer to remain a skeptic. People frequently misinterpret my opinion. I never said that every rich man is an idiot and every unsuccessful person unlucky, only that in absence of much additional information it is preferable to reserve one's judgment. It is safer.

•

Ten

LOSER TAKES ALL—
ON THE NONLINEARITIES OF LIFE

The nonlinear viciousness of life. Moving to Bel Air and acquiring the vices of the rich and famous. Why Microsoft's Bill Gates may not be the best in his business (but please do not inform him of such a fact). Depriving donkeys of food.

Next I put the platitude *life is unfair* under some examination, but from a new angle. The twist: Life is unfair in a *nonlinear* way. This chapter is about how a small advantage in life can translate into a highly disproportionate payoff, or, more viciously, how no advantage at all, but a very, very small help from randomness, can lead to a bonanza.

THE SANDPILE EFFECT

First we define *nonlinearity*. There are many ways to present it, but one of the most popular ones in science is what is called the sandpile effect, which I can illustrate as follows. I am currently sitting on a beach in Copacabana, in Rio de Janeiro, attempting to do nothing strenuous, away from anything to read and write (unsuccessfully, of course, as I am mentally writing these lines). I am playing with plastic beach toys borrowed from a child, trying to build an edifice—modestly but doggedly attempting to emulate the Tower of Babel. I continuously add sand to the top, slowly raising the entire structure. My Babylonian relatives thought they could thus reach the heavens. I have more humble designs—to test how high I can go before it topples. I keep adding sand, testing to see how the structure will ultimately collapse. Unused to seeing adults build sandcastles, a child looks at me with amazement.

In time—and much to the onlooking child's delight—my castle inevitably topples to rejoin the rest of the sand on the beach. It could be said that the last grain of sand is responsible for the destruction of the entire structure. What we are witnessing here is a nonlinear effect resulting from a linear force exerted on an object. A very small additional input, here the grain of sand, caused a disproportionate result, namely the destruction of my starter Tower of Babel. Popular wisdom has integrated many such phenomena, as witnessed by such expressions as "the straw that broke the camel's back" or "the drop that caused the water to spill."

These nonlinear dynamics have a bookstore name, "chaos theory," which is a misnomer because it has nothing to do with chaos. Chaos theory concerns itself primarily with functions in which a small input can lead to a disproportionate response. Population models, for instance, can lead to a path of explosive growth, or extinction of a species, depending on a very small difference in the population at a starting point in time. Another popular scientific analogy is the weather, where it has been shown that a simple butterfly fluttering its wings in India can cause a hurricane in New York. But the classics have their share to offer as well: Pascal (he of the wager in Chapter 7) said that if Cleopatra's nose had been slightly shorter, the world's fate would have changed. Cleopatra had comely features dominated by a thin and elongated nose that made Julius Caesar and his successor, Marc Antony, fall for her (here the intellectual snob in me cannot resist dissenting against conventional wisdom; Plutarch claimed that it was Cleopatra's skills in conversation, rather than her good looks, that caused the maddening infatuation of the shakers and movers of her day; I truly believe it).

Enter Randomness

Things can become more interesting when randomness enters the game. Imagine a waiting room full of actors queuing for an audition. The number of actors who will win is clearly small, and they are the ones generally observed by the public as representative of the profession, as we saw in our discussion on survivorship bias. The winners would move into Bel Air, feel pressure to acquire some basic training in the consumption of luxury goods, and, perhaps owing to the dissolute and unrhythmic lifestyle, flirt with substance abuse. As to the others (the great majority), we can imagine their fate; a lifetime of serving foamed *caffe latte* at the neighboring Starbucks, fighting the biological clock between auditions.

One may argue that the actor who lands the lead role that catapults him into fame and expensive swimming pools has some skills others lack, some charm, or a specific physical trait that is a perfect match for such a career path. I beg to differ. The winner may have some acting skills, but so do all of the others, otherwise they would not be in the waiting room.

It is an interesting attribute of fame that it has its own dynamics. An actor becomes known by some parts of the public because he is known by other parts of the public. The dynamics of such fame follow a rotating helix, which may have started at the audition, as the selection could have been caused by some silly detail that fitted the mood of the examiner on that day. Had the examiner not fallen in love the previous day with a person with a similar-sounding last name, then our selected actor from that particular sample *history* would be serving *caffe latte* in the intervening sample *history*.

Learning to Type

Researchers frequently use the example of QWERTY to describe the vicious dynamics of winning and losing in an economy, and to illustrate how the final outcome is more than frequently the undeserved one. The arrangement of the letters on a typewriter is an example of the success of the least deserving method. For our typewriters have the order of the letters on their keyboard arranged in a nonoptimal manner, as a matter of fact in such a nonoptimal manner as to slow down the typing rather than make the job easy, in order to avoid jamming the ribbons as they were designed for less electronic days. Therefore, as we started building better typewriters and computerized word processors, several attempts were made to rationalize the computer keyboard, to no avail. People were trained on a QWERTY keyboard and their habits were too sticky for change. Just like the helical propulsion of an actor into stardom, people patronize what other people like to do. Forcing rational dynamics on the process would be superfluous, nay, impossible. This is called a *path dependent outcome*, and has thwarted many mathematical attempts at modeling behavior.

It is obvious that the information age, by homogenizing our tastes, is causing the unfairness to be even more acute—those who win capture almost all the customers. The example that strikes many as the most spectacular lucky success is that of the software maker Microsoft and its moody founder Bill Gates. While it is hard to deny that Gates is a man of high personal standards, work ethics, and above-average intelligence, is he the best? Does he *deserve* it? Clearly not. Most people are equipped with his software (like myself) because other people are equipped with his software, a purely circular effect (economists call that "network externalities"). Nobody ever claimed that it was the best software product. Most of Gates' rivals have an obsessive jealousy of his success. They are maddened by the fact that he managed to win so big while many of them are struggling to make their companies survive.

Such ideas go against classical economic models, in which results either come from a precise reason (there is no account for uncertainty) or the good guy wins (the good guy is the one who is more skilled and has some technical superiority). Economists discovered path-dependent effects late in their game, then tried to publish wholesale on the topic that otherwise would be bland and obvious. For instance, Brian Arthur, an economist concerned with nonlinearities at the Santa Fe Institute, wrote that chance events coupled with positive feedback rather than technological superiority will determine economic superiority—not some abstrusely defined edge in a given area of expertise. While early economic models excluded randomness, Arthur explained how "unexpected orders, chance meetings with lawyers, managerial whims . . . would help determine which ones achieved early sales and, over time, which firms dominated."

MATHEMATICS INSIDE AND OUTSIDE THE REAL WORLD

A mathematical approach to the problem is in order. While in conventional models (such as the well-known Brownian random walk used in finance) the probability of success does not change with every incremental step, only the accumulated wealth, Arthur suggests models such as the Polya process, which is mathematically very difficult to work with, but can be easily understood with the aid of a Monte Carlo simulator. The Polya process can be presented as follows: Assume an urn initially containing equal quantities of black and red balls. You are to guess each time which color you will pull out before you make the draw. Here the game is rigged. Unlike a conventional urn, the probability of guessing correctly depends on past success, as you get better or worse at guessing depending on past performance. Thus, the probability of winning increases after past wins, that of losing increases after past losses. Simulating such a process, one can see a huge variance of outcomes, with astonishing successes and a large number of failures (what we called skewness).

Compare such a process with those that are more commonly modeled, that is, an urn from which the player makes guesses with replacement. Say you played roulette and won. Would this increase your chances of winning again? No. In a Polya process case, it does. Why is this so mathematically hard to work with? Because the notion of independence (i.e., when the next draw does not depend on past outcomes) is violated. Independence is a requirement for working with the (known) math of probability.

What has gone wrong with the development of economics as a science? Answer: There was a bunch of intelligent people who felt compelled to use mathematics just to tell themselves that they were rigorous in their thinking, that theirs was a science. Someone in a great rush decided to introduce mathematical modeling techniques (culprits: Leon Walras, Gerard Debreu, Paul Samuelson) without considering the fact that either the class of mathematics they were using was too restrictive for the class of problems they were dealing with, or that perhaps they should be aware that the precision of the language of mathematics could lead people to believe that they had solutions when in fact they had none (recall Popper and the costs of taking science too seriously). Indeed the mathematics they dealt with did not work in the real world, possibly because we needed richer classes of processes—and they refused to accept the fact that no mathematics at all was probably better.

The so-called *complexity theorists* came to the rescue. Much excitement was generated by the works of scientists who specialized in nonlinear quantitative methods—the mecca of those being the

Santa Fe Institute near Santa Fe, New Mexico. Clearly these scientists are trying hard, and providing us with wonderful solutions in the physical sciences and better models in the social siblings (though nothing satisfactory there yet). And if they ultimately do not succeed, it will simply be because mathematics may be of only secondary help in our real world. Note another advantage of Monte Carlo simulations is that we can get results where mathematics fails us and can be of no help. In freeing us from equations it frees us from the traps of inferior mathematics. As I said in Chapter 3, mathematics is merely a way of thinking and meditating, little more, in our world of randomness.

The Science of Networks

Studies of the dynamics of networks have mushroomed recently. They became popular with Malcolm Gladwell's book *The Tipping Point*, in which he shows how some of the behaviors of variables such as epidemics spread extremely fast beyond some unspecified critical level. (Like, say, the use of sneakers by inner-city kids or the diffusion of religious ideas. Book sales witness a similar effect, exploding once they cross a significant level of word-of-mouth.) Why do some ideologies or religions spread like wildfire while others become rapidly extinct? How do fads catch fire? How do idea viruses proliferate? Once one exits the conventional models of randomness (the bell curve family of charted randomness), something acute can happen. Why does the Internet hub Google get so many hits as compared to that of the National Association of Retired Veteran Chemical Engineers? The more connected a network, the higher the probability of someone hitting it and the more connected it will be, especially if there is no meaningful limitation on such capacity. Note that it is sometimes foolish to look for precise "critical points" as they may be unstable and impossible to know except, like many things, after the fact. Are these "critical points" not quite points but progressions (the so-called Pareto power laws)? While it is clear that the world produces clusters it is also sad that these may be too difficult to predict (outside of physics) for us to take their models seriously. Once again the important fact is knowing the existence of these nonlinearities, not trying to model them. The value of the great Benoit Mandelbrot's work lies more in telling us that there is a "wild" type of randomness of which we will never know much (owing to their unstable properties).

Our Brain

Our brain is not cut out for nonlinearities. People think that if, say, two variables are causally linked, then a steady input in one variable should *always* yield a result in the other one. Our emotional apparatus is designed for linear causality. For instance, you study every day and learn something in proportion to your studies. If you do not feel that you are going anywhere, your emotions will cause you to become demoralized. But reality rarely gives us the privilege of a satisfying linear positive progression: You may study for a year and learn nothing, then, unless you are disheartened by the empty results and give up, something will come to you in a flash. My partner Mark Spitznagel summarizes it as follows: Imagine yourself practicing the piano every day for a long time, barely being able to perform "Chopsticks," then suddenly finding yourself capable of playing Rachmaninov. Owing to this nonlinearity, people cannot comprehend the nature of the rare event. This summarizes why there are routes to success that are nonrandom, but few, very few, people have the mental stamina to follow them. Those who go the extra mile are rewarded. In my profession one may own a security that benefits from lower market prices, but may not react at all until some critical point. Most people give up before the rewards.

Buridan's Donkey or the Good Side of Randomness

Nonlinearity in random outcomes is sometimes used as a tool to break stalemates. Consider the problem of the nonlinear nudge.

Imagine a donkey equally hungry and thirsty placed at exactly equal distance from sources of food and water. In such a framework, he would die of both thirst and hunger as he would be unable to decide which one to get to first. Now inject some randomness in the picture, by randomly nudging the donkey, causing him to get closer to one source, no matter which, and accordingly away from the other. The impasse would be instantly broken and our happy donkey will be either in turn well fed then well hydrated, or well hydrated then well fed.

The reader no doubt has played a version of Buridan's donkey, by "flipping a coin" to break some of the minor stalemates in life where one lets randomness help with the decision process. Let Lady Fortuna make the decision and gladly submit. I often use Buridan's donkey (under its mathematical name) when my computer goes into a freeze between two possibilities (to be technical, these "randomizations" are frequently done during optimization problems, when one needs to perturbate a function).

Note that Buridan's donkey was named after the fourteenth-century philosopher Jean Buridan. Buridan had an interesting death (he was thrown in the Seine tied in a bag and died drowning). This tale was considered an example of sophistry by his contemporaries who missed the import of randomization—Buridan was clearly ahead of his time.

WHEN IT RAINS, IT POURS

As I am writing these lines, I am suddenly realizing that the world's bipolarity is hitting me very hard. Either one succeeds wildly, by attracting all the cash, or fails to draw a single penny. Likewise with books. Either everyone wants to publish it, or nobody is interested in returning telephone calls (in the latter case my discipline is to delete the name from my address book). I am also realizing the nonlinear effect behind success in anything: It is better to have a handful of enthusiastic advocates than hordes of people who appreciate your work—better to be loved by a dozen than liked by the hundreds. This applies to the sales of books, the spread of ideas, and success in general and runs counter to conventional logic. The information age is worsening this effect. This is making me, with my profound and antiquated Mediterranean sense of *metron* (measure), extremely uncomfortable, even queasy. Too much success is the enemy (think of the punishment meted out on the rich and famous); too much failure is demoralizing. I would like the option of having neither.

•

Eleven

RANDOMNESS AND OUR MIND: WE ARE PROBABILITY BLIND

On the difficulty of thinking of your vacation as a linear combination of Paris and the Bahamas. Nero Tulip may never ski in the Alps again. Do not ask bureaucrats too many questions. A Brain Made in Brooklyn. We need Napoleon. Scientists bowing to the King of Sweden. A little more on journalistic pollution. Why you may be dead by now.

PARIS OR THE BAHAMAS?

You have two options for your next brief vacation in March. The first is to fly to Paris; the second is to go to the Caribbean. You expressed indifference between the two options; your spouse will tip the decision one way or the other. Two distinct and separate images come to you when you think of the possibilities. In the first one, you see yourself standing at the Musée d'Orsay in front of some Pissaro painting depicting a cloudy sky—the gray Parisian wintry sky. You are carrying an umbrella under your arm. In the second image, you are lying on a towel with a stack of books by your favorite authors next to you (Tom Clancy and Ammianus Marcellinus), and an obsequious waiter serving you a banana daiquiri. You know that the two states are mutually exclusive (you can only be in one place at one time), but exhaustive (there is a 100% probability that you will be in one of them). They are equiprobable, with, in your opinion, 50% probability assigned to each.

You derive great pleasure thinking about your vacation; it motivates you and makes your daily commute more bearable. But the adequate way to visualize yourself, according to rational behavior under uncertainty, is 50% in one of the vacation spots and 50% in the other—what is mathematically called a *linear combination* of the two states. Can your brain handle that? How desirable would it be to have your feet in the Caribbean waters and your head exposed to the Parisian rain? Our brain can properly handle one and only one state at once—unless you have personality troubles of a deeply pathological nature. Now try to imagine an 85%/15% combination. Any luck?

Consider a bet you make with a colleague for the amount of $1,000, which, in your opinion, is exactly fair. Tomorrow night you will have zero or $2,000 in your pocket, each with a 50% probability. In purely mathematical terms, the fair value of a bet is the linear combination of the states, here called the *mathematical expectation*, i.e., the probabilities of each payoff multiplied by the dollar values at stake (50% multiplied by 0 and 50% multiplied by $2,000 = $1,000). Can you *imagine* (that is visualize, not compute mathematically) the value being $1,000? We can conjure up one *and only one* state at a given time, i.e., either 0 or $2,000. Left to our own devices, we are likely to bet in an irrational way, as one of the states would dominate the picture—the fear of ending with nothing or the excitement of an extra $1,000.

SOME ARCHITECTURAL CONSIDERATIONS

Time to reveal Nero's secret. It was a black swan. He was then thirty-five. Although prewar buildings in New York can have a pleasant front, their architecture seen from the back offers a stark contrast by being completely bland. The doctor's examination room had a window overlooking the backyard of one such Upper East Side street, and Nero will always remember how bland that backyard was in comparison with the front, even if he were to live another half century. He will always remember the view of the ugly pink backyard from the leaden window panes, and the medical diploma on the wall that he read a dozen times as he was waiting for the doctor to come into the room (half an eternity, for Nero suspected that something was wrong). The news was then delivered (grave voice), "I have some . . . I got the pathology report . . . It's . . . It is not as bad as it sounds . . . It's . . . It's cancer." The declaration caused his body to be hit by an electric discharge, running through his back down to his knees. Nero tried to yell "What?" but no sound came out of his mouth. What scared him was not so much the news as the sight of the doctor. Somehow the news reached his body before his mind. There was too much fear in the doctor's eyes and Nero immediately suspected that the news was even worse than what he was being told (it was).

The night of the diagnosis, at the medical library where he sat, drenched wet from walking for hours in the rain without noticing it and making a puddle of water around him (he was yelled at by an attendant but could not concentrate on what she was saying so she shrugged her shoulders and walked away); later he read the sentence "72% 5-year actuarially adjusted survival rate." It meant that 72 people out of 100 make it. It takes between three and five years for the body without clinical manifestations of the disease for the patient to be pronounced cured (closer to three at his age). He then felt in his guts quite certain that he was going to make it.

Now the reader might wonder about the mathematical difference between a 28% chance of death and a 72% chance of survival over the next five years. Clearly, there is none, but we are not made for mathematics. In Nero's mind a 28% chance of death meant the image of himself dead, and thoughts of the cumbersome details of his funeral. A 72% chance of survival put him in a cheerful mood; his mind was planning the result of a cured Nero skiing in the Alps. At no point during his ordeal did Nero think of himself as 72% alive and 28% dead.

Just as Nero cannot "think" in complicated shades, consumers consider a 75% fat-free hamburger to be different from a 25% fat one. Likewise with statistical significance. Even specialists tend to infer too fast from data in accepting or rejecting things. Recall the dentist whose emotional well-being depends on the recent performance of his portfolio. Why? Because as we will see, rule-determined behavior does not require nuances. Either you kill your neighbor or you don't. Intermediate sentiments (leading, say, to only half his killing) are either useless or downright dangerous when you do things. The emotional apparatus that jolts us into action does not understand such nuances—it is not efficient to understand things. The rest of this chapter will rapidly illustrate some manifestations of such blindness, with a cursory exposition of the research in that area (only what connects to the topics in this book).

BEWARE THE PHILOSOPHER BUREAUCRAT

For a long time we had the wrong product specifications when we thought of ourselves. We humans have been under the belief that we were endowed with a beautiful machine for thinking and understanding things. However, among the factory specifications for us is the lack of awareness of the true factory specifications (why complicate things?). The problem with thinking is that it causes you to develop illusions. And thinking may be such a waste of energy! Who needs it!

Consider that you are standing in front of a government clerk in a heavily socialist country where being a bureaucrat is held to be what respectable people do for a living. You are there to get your papers stamped by him so you can export some of their lovely chocolate candies to the New Jersey area, where you think the local population would have a great taste for them. What do you think his function is? Do you think for a minute that he cares about the general economic theory behind the transaction? His job is just to verify that you have the twelve or so signatures from the right departments, true/false; then stamp your papers and let you go. General considerations of economic growth or balance of trade are none of his interests. In fact you are lucky that he doesn't spend any time meditating about these things: Consider how long the procedure would take if he had to solve balance of trade equations. He just has a rulebook and, over a career spanning forty to forty-five years, he will just stamp documents, be mildly rude, and go home to drink nonpasteurized beer and watch soccer games. If you gave him Paul Krugman's book on international economics he would either sell it in the black market or give it to his nephew.

Accordingly, rules have their value. We just follow them not because they are the best but because they are useful and they save time and effort. Consider that those who started theorizing upon seeing a tiger on whether the tiger was of this or that taxonomic variety, and the degree of danger it represented, ended up being eaten by it. Others who just ran away at the smallest presumption and were not slowed down by the smallest amount of thinking ended up either outchasing the tiger or outchasing their cousin who ended up being eaten by it.

Satisficing

It is a fact that our brains would not be able to operate without such shortcuts. The first thinker who figured it out was Herbert Simon, an interesting fellow in intellectual history. He started out as a political scientist (but he was a formal thinker, not the literary variety of political scientists who write about Afghanistan in *Foreign Affairs*); he was an artificial-intelligence pioneer, taught computer science and psychology, did research in cognitive science, philosophy, and applied mathematics, and received the Bank of Sweden Prize for Economics in honor of Alfred Nobel. His idea is that if we were to optimize at every step in life, then it would cost us an infinite amount of time and energy. Accordingly, there has to be in us an approximation process that stops somewhere. Clearly he got his intuitions from computer science—he spent his entire career at Carnegie-Mellon University in Pittsburgh, which has a reputation as a computer science center. "Satisficing" was his idea (the melding together of *satisfy* and *suffice*): You stop when you get a near-satisfactory solution. Otherwise it may take you an eternity to reach the smallest conclusion or perform the smallest act. We are therefore rational, but in a limited way: "boundedly rational." He believed that our brains were a large optimizing machine that had built-in rules to stop somewhere.

Not quite so, perhaps. It may not be just a rough approximation. For two (initially) Israeli researchers on human nature, how we behave seemed to be a completely different process from the optimizing machine presented by Simon. The two sat down introspecting in Jerusalem looking at aspects of their own thinking, compared it to rational models, and noticed *qualitative* differences. Whenever they both seemed to make the same mistake of reasoning they ran empirical tests on subjects, mostly students, and discovered very surprising results on the relation between thinking and rationality. It is to their discovery that we turn next.

FLAWED, NOT JUST IMPERFECT

Kahneman and Tversky

Who has exerted the most influence on economic thinking over the past two centuries? No, it is not John Maynard Keynes, not Alfred Marshall, not Paul Samuelson, and certainly not Milton Friedman. The answer is two noneconomists: Daniel Kahneman and Amos Tversky, the two Israeli introspectors, and their specialty was to uncover areas where human beings are not endowed with rational probabilistic thinking and optimal behavior under uncertainty. Strangely, economists studied uncertainty for a long time and did not figure out much—if anything, they thought they knew something and were fooled by it. Aside from some penetrating minds like Keynes, Knight, and Shackle, economists did not even figure out that they had no clue about uncertainty—the discussions on risk by their idols show that *they did not know how much they did not know*. Psychologists, on the other hand, looked at the problem and came out with solid results. Note that, unlike economists, they conducted experiments, true controlled experiments of a repeatable nature, that can be done in Ulan Bator, Mongolia, tomorrow if necessary. Conventional economists do not have this luxury as they observe the past and make lengthy and mathematical comments, then bicker with each other about them.

Kahneman and Tversky went in a completely different direction than Simon and started figuring out rules in humans that did not make them rational—but things went beyond the shortcut. For them, these rules, which are called *heuristics*, were not merely a simplification of rational models, but were different in methodology and category. They called them "quick and dirty" heuristics. There is a dirty part: These shortcuts came with side effects, these effects being the biases, most of which I discussed previously throughout the text (such as the inability to accept anything abstract as risk). This started an empirical research tradition called the "heuristics and biases" tradition that attempted to catalogue them—it is impressive because of its empiricism and the experimental aspect of the methods used.

Since the Kahneman and Tversky results, an entire discipline called behavioral finance and economics has flourished. It is in open contradiction with the orthodox so-called neoclassical economics taught in business schools and economics departments under the normative names of efficient markets, rational expectations, and other such concepts. It is worth stopping, at this juncture, and discussing the distinction between normative and positive sciences. A normative science (clearly a self-contradictory concept) offers prescriptive teachings; it studies how things *should* be. Some economists, for example those of the efficient-market religion, believe that our studies should be based on the hypothesis that humans are rational and act rationally because it is the best thing for them to do (it is mathematically "optimal"). The opposite is a positive science, which is based on how people actually are observed to behave. In spite of economists' envy of physicists, physics is an inherently positive science while economics, particularly microeconomics and financial economics, is predominantly a normative one. Normative economics is like religion without the aesthetics.

Note that the experimental aspect of the research implies that Daniel Kahneman and the experimental ponytailed economist Vernon Smith were the first true scientists ever to bow in front of the Swedish king for the economics prize, something that should give credibility to the Nobel academy, particularly if, like many, one takes Daniel Kahneman far more seriously than a collection of serious-looking (and very human, hence fallible) Swedes. There is another hint of the scientific firmness of this research: It is extremely readable for someone outside of psychology, unlike papers in conventional economics and finance that even people in the field have difficulty reading (as the discussions are jargon-laden and heavily mathematical to give the illusion of science). A motivated reader can get concentrated in four volumes the collection of the major heuristics and biases papers.

Economists were not at the time very interested in hearing these stories of irrationality: *Homo economicus* as we said is a normative concept. While they could easily buy the "Simon" argument that we are not perfectly rational and that life implies approximations, particularly when the stakes are not large enough, they were not willing to accept that people were flawed rather than imperfect. But they are. Kahneman and Tversky showed that these biases do not disappear when there are incentives, which means that they are not necessarily cost saving. They were a different form of reasoning, and one where the probabilistic reasoning was weak.

WHERE IS NAPOLEON WHEN WE NEED HIM?

If your mind operates by series of different disconnected rules, these may not be necessarily consistent with each other, and if they may still do the job *locally*, they will not necessarily do so *globally*. Consider them stored as a rulebook of sorts. Your reaction will depend on which page of the book you open to at any point in time. I will illustrate it with another socialist example.

After the collapse of the Soviet Union, Western businesspeople involved in what became Russia discovered an annoying (or entertaining) fact about the legal system: It had conflicting and contradictory laws. It just depended on which chapter you looked up. I don't know whether the Russians wanted it as a prank (after all, they lived long, humorless years of oppression) but the confusion led to situations where someone had to violate a law to comply with another. I have to say that lawyers are quite dull people to talk to; talking to a dull lawyer who speaks broken English with a strong accent and vodka breath can be quite straining—so you give up. This spaghetti legal system came from the piecewise development of the rules: You add a law here and there and the situation is too complicated as there is no central system that is consulted every time to ensure compatibility of all the parts together. Napoleon faced a similar situation in France and remedied it by setting up a top-down code of law that aimed to dictate a full logical consistency. The problem with us humans is not so much that no Napoleon has showed up so far to dynamite the old structure then reengineer our minds like a big central program; it is that our minds are far more complicated than just a system of laws, and the requirement for efficiency is far greater.

Consider that your brain reacts differently to the same situation depending on which chapter you open to. The absence of a central processing system makes us engage in decisions that can be in conflict with each other. You may prefer apples to oranges, oranges to pears, but pears to apples—it depends on how the choices are presented to you. The fact that your mind cannot retain and use everything you know at once is the cause of such biases. One central aspect of a heuristic is that it is blind to reasoning.

"I'm As Good As My Last Trade" and Other Heuristics

There exist plenty of different catalogues of these heuristics in the literature (many of them overlapping); the object of this discussion is to provide the intuition behind their formation rather than list them. For a long time we traders were totally ignorant of the behavioral research and saw situations where there was with strange regularity a wedge between the simple probabilistic reasoning and people's perception of things. We gave them names such as the "I'm as good as my last trade" effect, the "sound-bite effect," the "Monday morning quarterback" heuristic, and the "It was obvious after the fact" effect. It was both vindicating for traders' pride and disappointing to discover that they existed in the heuristics literature as the "anchoring," the "affect heuristic," and the "hindsight bias" (it makes us feel that trading is true, experimental scientific research). The correspondence between the two worlds is shown in Table 11.1.

I start with the "I'm as good as my last trade" heuristic (or the "loss of perspective" bias)—the fact that the counter is reset at zero and you start a new day or month from scratch, whether it is your accountant who does it or your own mind. This is the most significant distortion and the one that carries the most consequences. In order to be able to put things in general context, you

Table 11.1 Trader and Scientific Approach

Trader Name	Learned Name	Description
"I'm as good as my last trade."	Prospect theory	Looking at differences, not absolutes, and resetting to a specific reference point
"Sound-bite effect" or "Fade the fears"	Affect heuristic, risk-as-feeling theory	People react to concrete and visible risks, not abstract ones
"It was so obvious" or "Monday morning quarterback"	Hindsight bias	Things appear to be more predictable after the fact
"You were wrong"	Belief in the law of small numbers	Inductive fallacies; jumping to general conclusions too quickly
Brooklyn smarts/MIT intelligence	Two systems of reasoning	The working brain is not quite the reasoning one
"It will *never* go there"	Overconfidence	Risk-taking out of an underestimation of the odds

do not have everything you know in your mind at all times, so you retrieve the knowledge that you require at any given time in a piecemeal fashion, which puts these retrieved knowledge chunks in their local context. This means that you have an arbitrary reference point and react to differences from that point, forgetting that you are only looking at the differences from that particular perspective of the local context, not the absolutes.

There is the well-known trader maxim "life is incremental." Consider that as an investor you examine your performance like the dentist in Chapter 3, at some set interval. What do you look at: your monthly, your daily, your life-to-date, or your hourly per-

formance? You can have a good month and a bad day. Which period should dominate?

When you take a gamble, do you say: "My net worth will end
up at $99,000 or $101,500 after the gamble" or do you say "I lose
$1,000 or make $1,500?" Your attitude toward the risks and rewards of the gamble will vary according to whether you look at
your net worth or changes in it. But in fact in real life you will be
put in situations where you will only look at your *changes*. The fact
that the losses hurt more than the gains, and *differently*, makes
your accumulated performance, that is, your total wealth, less relevant than the last change in it.

This dependence on the local rather than the global status
(coupled with the effect of the losses hitting harder than the
gains) has an impact on your perception of well-being. Say you get
a windfall profit of $1 million. The next month you lose $300,000.
You adjust to a given wealth (unless of course you are very poor)
so the following loss would hurt you emotionally, something that
would not have taken place if you received the net amount of
$700,000 in one block, or, better, two sums of $350,000 each. In
addition, it is easier for your brain to detect differences rather than
absolutes, hence rich or poor will be (above the minimum level)
in relation to something else (remember Marc and Janet). Now,
when something is *in relation* to something else, that something
else can be manipulated. Psychologists call this effect of comparing to a given reference *anchoring*. If we take it to its logical limit
we would realize that, because of this resetting, wealth itself does
not really make one happy (above, of course, some subsistence
level); but positive changes in wealth may, especially if they come
as "steady" increases. More on that later with my discussion of option blindness.

Other aspects of anchoring. Given that you may use two different anchors in the same situation, the way you act depends on
so little. When people are asked to estimate a number, they will

position it with respect to a number they have in mind or one they just heard, so "big" or "small" will be comparative. Kahneman and Tversky asked subjects to estimate the proportion of African countries in the United Nations after making them consciously pull a random number between 0 and 100 (they knew it was a random number). People guessed in relation to that number, which they used as anchor: Those who randomized a high number guessed higher than those who randomized a low one. This morning I did my bit of anecdotal empiricism and asked the hotel concierge how long it takes to go to the airport. "40 minutes?" I asked. "About 35," he answered. Then I asked the lady at the reception if the journey was 20 minutes. "No, about 25," she answered. I timed the trip: 31 minutes.

This anchoring to a number is the reason people do not react to their total accumulated wealth, but to differences of wealth from whatever number they are currently anchored to. This is the major conflict with economic theory, as according to economists, someone with $1 million in the bank would be more satisfied than if he had half a million. But we saw John reaching $1 million having had a total of $10 million; he was happier when he only had half a million (starting at nothing) than where we left him in Chapter 1. Also recall the dentist whose emotions depended on how frequently he checked his portfolio.

Degree in a Fortune Cookie

I used to attend a health club in the middle of the day and chat with an interesting Eastern European fellow with two Ph.D. degrees, one in physics (statistical no less), the other in finance. He worked for a trading house and was obsessed with the anecdotal aspects of the markets. He once asked me doggedly what I thought the stock market would do that day. Clearly I gave him a social answer of the kind "I don't know, perhaps lower"—quite possibly the opposite answer to what I would have given him had he asked me an hour earlier. The next day he showed great alarm upon seeing me. He went on and on discussing my credibility and wondering how I could be so wrong in my "predictions," since the market went up subsequently. The man was able to derive conclusions about my ability to predict and my "credibility" with a single observation. Now, if I went to the phone and called him and disguised my voice and said, "Hello, this is Doktorr Talebski from the Academy of Lodz and I have an interrresting prrroblem," then presented the issue as a statistical puzzle, he would laugh at me. "Doktorr Talevski, did you get your degree in a fortune cookie?" Why is it so?

Clearly there are two problems. First, the quant did not use his statistical brain when making the inference, but a different one. Second, he made the mistake of overstating the importance of small samples (in this case just one single observation, the worst possible inferential mistake a person can make). Mathematicians tend to make egregious mathematical mistakes outside of their theoretical habitat. When Tversky and Kahneman sampled mathematical psychologists, some of whom were authors of statistical textbooks, they were puzzled by their errors. "Respondents put too much confidence in the result of small samples and their statistical judgment showed little sensitivity to sample size." The puzzling aspect is that not only *should* they have known better, "they *did* know better." And yet . . .

I will next list a few more heuristics. (1) The *availability* heuristic, which we saw in Chapter 3 with the earthquake in California deemed more likely than catastrophe in the entire country, or death from terrorism being more "likely" than death from all possible sources (including terrorism). It corresponds to the practice of estimating the frequency of an event according to the ease with which instances of the event can be recalled. (2) The *representativeness* heuristic: gauging the probability that a person belongs to a particular social group by assessing how similar the person's characteristics are to the "typical" group member's. A feminist-style philosophy student is deemed more likely to be a feminist bank teller than to be just a bank teller. This problem is known as the "Linda problem" (the feminist's name was Linda) and has caused plenty of academic ink to flow (some of the people engaged in the "rationality debate" believe that Kahneman and Tversky are putting highly normative demands on us humans). (3) The *simulation* heuristic: the ease of mentally undoing an event—playing the alternative scenario. It corresponds to counterfactual thinking: Imagine what might have happened had you not missed your train (or how rich you'd be today had you liquidated your portfolio at the height of the NASDAQ bubble). (4) We discussed in Chapter 3 the *affect* heuristic: What emotions are elicited by events determine their probability in your mind.

Two Systems of Reasoning

Later research refines the problem as follows: There are two possible ways for us to reason, the heuristics being part of one—rationality being part of the other. Recall the colleague who used a different brain in the classroom than the one in real life in Chapter 2. Didn't you wonder why the person you think knows physics so well cannot apply the basic laws of physics by driving well? Researchers divide the activities of our mind into the following two polarized parts, called System 1 and System 2.

System 1 is effortless, automatic, associative, rapid, parallel process, opaque (i.e., we are not aware of using it), emotional, concrete, specific, social, and personalized.

System 2 is effortful, controlled, deductive, slow, serial, self-aware, neutral, abstract, sets, asocial, and depersonalized.

I have always believed that professional option traders and market makers by dint of practicing their probabilistic game build an innate probabilistic machine that is far more developed than the rest of the population—even that of probabilists. I found a confirmation of that as researchers in the heuristics and biases tradition believe that System 1 can be impacted by experience and integrate elements from System 2. For instance, when you learn to play chess, you use System 2. After a while things become intuitive and you are able to gauge the relative strength of an opponent by glancing at the board.

Next I introduce the evolutionary psychology point of view.

WHY WE DON'T MARRY THE FIRST DATE

Another branch of research, called evolutionary psychology, developed a completely different approach to the same problem. It operates in parallel, creating some bitter but not too worrisome academic debates. These evolutionary psychologists agree with the Kahneman-Tversky school that people have difficulties with standard probabilistic reasoning. However, they believe that the reason lies in the way things are presented to us in the current environment. To them, we are optimized for a set of probabilistic reasoning, but in a different environment than the one prevailing today. The statement "Our brains are made for fitness not for truth" by the scientific intellectual Steven Pinker, the public spokesmen of that school, summarizes it all. They agree that our brains are not made for understanding things but think that they are not biased, or only biased because we do not use them in their real habitat.

Strangely, the Kahneman-Tversky school of researchers did not incur any credible resistance from the opinions of the economists of the time (the general credibility of conventional economists has always been so low that almost nobody in science or in the real world ever pays attention to them). No, instead the challenge came from the sociobiologists—and the center of the disagreement lies in their belief in using evolutionary theory as a backbone for our understanding of human nature. While this caused a fierce

scientific dispute, I will have to say that they agree on the significant part as far as this book is concerned: (1) We do not *think*
when making choices but use heuristics; (2) We make serious
probabilistic mistakes in today's world—*whatever the true reason*.
Note that the split even covers the new economics: Just as we
have a scientific branch of economics coming out of the Kahneman and Tversky tradition (behavioral economics), there is another scientific branch of economics coming out of evolutionary
psychology, with the caveman economics approach followed by
such researchers as the economist-biologist Terry Burnham, coau-
thor of the very readable *Mean Genes*.

Our Natural Habitat

I will not delve too deeply into amateur evolutionary theory to
probe at the reasons (besides, in spite of having spent some time in
libraries I feel that I am truly an amateur in the subject matter).
Clearly, the environment for which we have built our endowment
is not the one that prevails today. I have not told too many of my
colleagues that their decision making contains some lingering
habits of cavemen—but when markets experience an abrupt
move, I experience the same rush of adrenaline as if a leopard
were seen prowling near my trading desk. Some of my colleagues
who break telephone handles upon losing money might be even
closer in their psychological makeup to our common origin.

This might be a platitude to those who frequent the Greek and
Latin classics, but we never fail to be surprised when noticing that
people a couple of dozen centuries removed from us can exhibit
similar sensibility and feelings. What used to strike me as a child
upon visiting museums is that ancient Greek statues exhibit men
with traits indistinguishable from ours (only more harmonious
and aristocratic). I was so wrong to believe that 2,200 years was a
long time. Proust wrote frequently about the surprise people have
when coming across emotions in Homeric heroes that are similar

to those we experience today. By genetic standards, these Homeric heroes of thirty centuries ago in all likelihood have the exact identical makeup as the pudgy middle-aged man you see schlepping groceries in the parking lot. More than that. In fact, we are truly identical to the man who perhaps eighty centuries ago started being called "civilized," in that strip of land stretching from southeastern Syria to southwestern Mesopotamia.

What is our natural habitat? By natural habitat, I mean the environment in which we reproduced the most, the one in which we spent the highest number of generations. The consensus among anthropologists is that we have been around as a separate species for 130,000 years, most of which were spent in the African savannah. But we do not have to go back that far in history to get the point. Imagine life in an early urban settlement, in Middle-Town, Fertile Crescent, only about 3,000 years ago—surely modern times from a genetic standpoint. Information is limited by the physical means of its transmission; one cannot travel fast, hence information will come from faraway places in concise batches. Traveling is a nuisance fraught with all manner of physical danger; you will settle within a narrow radius of where you were born unless famine or some invading uncivilized tribe dislodges you and your relatives from your happy settlement. The number of people you would get to know in a lifetime will be small. Should a crime be committed, it will be easy to gauge the evidence of guilt within the small number of possible suspects. If you are unjustly convicted of a crime, you will argue in simple terms, propounding simple evidence like "I was not there as I was praying in the temple of Baal and was seen at dusk by the high priest" and add that Obedshemesh, son of Sahar, was more likely to be guilty because he had more to gain from the crime. Your life would be simple, hence your space of *probabilities* would be narrow.

The real problem is, as I have mentioned, that such a natural habitat does not include much information. An efficient computation of the odds was never necessary until very recently. This also explains why we had to wait until the emergence of the gambling literature to see the growth of the mathematics of probability. Popular belief holds that the religious backdrop of the first and second millennia blocked the growth of tools that hint at absence of determinism, and caused the delays in probability research. The idea is extremely dubious; we simply did not compute probabilities because we did not *dare* to? Surely the reason is rather because we did not *need* to. Much of our problem comes from the fact that we have evolved out of such a habitat faster, much faster, than our genes. Even worse, our genes have not changed at all.

Fast and Frugal

Evolutionary theorists agree that brainwork depends on how the subject is presented and the frame offered—and they can be contradictory in their results. We detect cheaters with a different part of our brain than the one we draw on to solve logical problems. People can make incoherent choices because the brain works in the form of small partial jobs. Those heuristics that we said were "quick and dirty" to the psychologists are "fast and frugal" to the evolutionary psychologists. Not only that, but some thinkers, like the cognitive scientist Gerd Gigerenzer, seem to have obsessively taken the other side of the trade from Kahneman and Tversky; his work and that of his associates at the ABC Group (Adaptive Behavior and Cognition) intend to show that we are rational and that evolution produces a form of rationality he calls "ecological rationality." They believe that not only are we hard-wired for *optimizing probabilistic* behavior in situations like mate selection (how many people of the opposite sex do you need to meet before pulling the trigger?), or choosing a meal, but we are also so wired for stock selection and that we do it appropriately if the stocks are presented to us in the correct manner.

In fact, Gigerenzer agrees that we do not understand probability (too abstract), but we react rather well to frequencies (less abstract): According to him, some problems that normally would cause us to make a mistake disappear when phrased in terms of percentages.

According to these researchers, while we may like to think of our brain as a central processing system, with top-down features, an analogy to the Swiss Army knife (with its small specific tools) seems to be in order. How? The psychologists' framework is built around the distinction between the domain-specific and domain-general adaptations. A domain-specific adaptation is something that is meant to solve a very precise task (as opposed to domain-general ones that are meant to solve global ones). While these are easy to understand and accept for physiological adaptations (i.e., a giraffe's neck helps in reaching food or an animal's colors in providing camouflage), people have had difficulties accepting why these apply to our mind in the same manner.

Our brain functions by "modules." An interesting aspect of modularity is that we may use different modules for different instances of the *same* problem, depending on the framework in which it is presented—as discussed in the notes to this section. One of the attributes of a module is its "encapsulation," i.e., we cannot interfere with its functioning, as we are not aware of using it. The most striking module is used when we try to find a cheater. Expressed in purely logical form (though with extreme clarity), a given quiz is only solved by 15% of the people to whom it is given. Now, the same quiz expressed in a manner that aims at uncovering a cheater, almost everyone gets it.

Neurobiologists Too

Neurobiologists also have their side of the story. They believe (roughly) that we have three brains: The very old one, the reptilian brain that dictates heartbeat and that we share with all animals; the limbic brain center of emotions that we share with mammals; and the neocortex, or cognitive brain, that distinguishes primates and humans (note that even institutional investors seem to have a neocortex). While that theory of the Triune brain shows some oversimplification (particularly when handled by journalists), it seems to provide a framework for the analysis of brain functions.

Although it is very difficult to figure out which part of the brain does what exactly, neuroscientists have been doing some environment mapping in the brain by, say, taking a patient whose brain is damaged in one single spot (say, by a tumor or an injury deemed to be local) and deducing by elimination the function performed by such part of the anatomy. Other methods include brain imaging and electric simulations to specific areas. Many researchers outside of neurobiology, like the philosopher and cognitive scientist Jerry Fodor (who pioneered the notion of modularity) remain skeptical about the quality of the knowledge that we can uncover by examining the physical properties of the brain, be it only on account of the complicated interactions of the single parts (with corresponding nonlinearities). The mathematician and cognitive scientist David Marr, who pioneered the field of object recognition, made the apt remark that one does not learn how birds fly by studying feathers but rather by studying aerodynamics. I will present the theses of two watershed works presented in readable books, Damasio's *Descartes' Error* and LeDoux's *Emotional Brain*.

Descartes' Error presents a very simple thesis: You perform a surgical ablation on a piece of someone's brain (say, to remove a tumor and tissue around it) with the sole resulting effect of an inability to register emotions, nothing else (the IQ and every other faculty remain the same). What you have done is a controlled experiment to separate someone's intelligence from his emotions. Now you have a purely rational human being unencumbered with feelings and emotions. Let's watch: Damasio reported that the purely unemotional man was incapable of making the simplest decision. He could not get out of bed in the morning, and frittered away his days fruitlessly weighing decisions. Shock! This flies in the face of everything one would have expected: One cannot make a decision without emotion. Now, mathematics gives the same answer: If one were to perform an optimizing operation across a large collection of variables, even with a brain as large as ours, it would take a very long time to decide on the simplest of tasks. So we need a shortcut; emotions are there to prevent us from temporizing. Does it remind you of Herbert Simon's idea? It seems that the emotions are the ones doing the job. Psychologists call them "lubricants of reason."

Joseph LeDoux's theory about the role of emotions in behavior is even more potent: Emotions affect one's thinking. He figured out that much of the connections from the emotional systems to the cognitive systems are stronger than connections from the cognitive systems to the emotional systems. The implication is that we feel emotions (limbic brain) then find an explanation (neocortex). As we saw with Claparède's discovery, much of the opinions and assessments that we have concerning risks may be the simple result of emotions.

Kafka in a Courtroom

The O. J. Simpson trial provides an example of how our modern society is ruled by probability (because of the explosion in information), while important decisions are made without the smallest regard for its basic laws. We are capable of sending a spacecraft to Mars, but we are incapable of having criminal trials managed by the basic laws of probability—yet evidence is clearly a probabilistic notion. I remember buying a book on probability at a Borders Books chain bookstore only a short distance from the Los Angeles courthouse where the "trial of the century" was taking place—another book that crystallized the highly sophisticated quantitative knowledge in the field. How could such a leap in knowledge elude lawyers and jurors only a few miles away?

People who are as close to being criminal as probability laws can allow us to infer (that is, with a confidence that exceeds the *shadow of a doubt*) are walking free because of our misunderstanding of basic concepts of the odds. Equally, you could be convicted for a crime you never committed, again owing to a poor reading of probability—for we still cannot have a court of law properly compute the joint probability of events (the probability of two events taking place at the same time). I was in a dealing room with a TV set turned on when I saw one of the lawyers arguing that there were at least four people in Los Angeles capable of carrying O. J. Simpson's DNA characteristics (thus ignoring the joint set of events—we will see how in the next paragraph). I then switched off the television set in disgust, causing an uproar among the traders. I was under the impression until then that sophistry had been eliminated from legal cases thanks to the high standards of republican Rome. Worse, one Harvard lawyer used the specious argument that only 10% of men who brutalize their wives go on to murder them, which is a probability unconditional on the murder (whether the statement was made out of a warped notion of advocacy, pure malice, or ignorance is immaterial). Isn't the law devoted to the truth? The correct way to look at it is to determine the percentage of murder cases where women were killed by their husbands *and* had previously been battered by them (that is, 50%)—for we are dealing with what is called *conditional* probabilities; the probability that O.J. killed his wife *conditional* on the information of her having been killed, rather than the *unconditional* probability of O.J. killing his wife. How can we expect the untrained person to understand randomness when a Harvard professor who deals and teaches the concept of probabilistic evidence can make such an incorrect statement?

More particularly, where jurors (and lawyers) tend to make mistakes, along with the rest of us, is in the notion of joint probability. They do not realize that evidence compounds. The probability of my being diagnosed with respiratory tract cancer and being run over by a pink Cadillac in the same year, assuming each one of them is 1/100,000, becomes 1/10,000,000,000—by multiplying the two (obviously independent) events. Arguing that O. J. Simpson had 1/500,000 chance of not being the killer from the blood standpoint (remember the lawyers used the sophistry that there were four people with such blood types walking around Los Angeles) and adding to it the fact that he was the husband of the person and that there was additional evidence, then (owing to the compounding effect) the odds against him rise to several trillion trillion.

"Sophisticated" people make worse mistakes. I can surprise people by saying that the probability of the joint event is lower than either. Recall the availability heuristic: with the Linda problem rational and educated people finding the likelihood of an event greater than that of a larger one that encompasses it. I am glad to be a trader taking advantage of people's biases but I am scared of living in such a society.

An Absurd World

Kafka's prophetic book, *The Trial*, about the plight of a man, Joseph K., who is arrested for a mysterious and unexplained reason, hit a spot as it was written before we heard of the methods of the "scientific" totalitarian regimes. It projected a scary future of mankind wrapped in absurd self-feeding bureaucracies, with spontaneously emerging rules subjected to the internal logic of the bureaucracy. It spawned an entire "literature of the absurd"; the world may be too incongruous for us. I am terrified of certain lawyers. After listening to statements during the O.J. trial (and their effect) I was scared, truly scared, of the possible outcome—my being arrested for some reason that made no sense probabilistically, and having to fight some glib lawyer in front of a randomness illiterate jury.

We said that mere judgment would probably suffice in a primitive society. It is easy for a society to live without mathematics—or traders to trade without quantitative methods—when the space of possible outcomes is one-dimensional. One-dimensional means that we are looking at one sole variable, not a collection of separate events. The price of one security is one-dimensional, whereas the collection of the prices of several securities is multi-dimensional and requires mathematical modeling—we cannot easily see the collection of possible outcomes of the portfolio with a naked eye, and cannot even represent it on a graph as our physical world has been limited to visual representation in three dimensions only. We will argue later why we run the risk of having bad models (admittedly, we have) or making the error of condoning ignorance—swinging between the Carybde of the lawyer who knows no math to the Scylla of the mathematician who misuses his math because he does not have the judgment to select the right model. In other words, we will have to swing between the mistake of listening to the glib nonsense of a lawyer who refuses science and that of applying the flawed theories of some economist who takes his science too seriously. The beauty of science is that it makes an allowance for both error types. Luckily, there is a middle road—but sadly, it is rarely traveled.

Examples of Biases in Understanding Probability

I found in the behavioral literature at least forty damning examples of such acute biases, systematic departures from rational behavior widespread across professions and fields. Below is the account of a well-known test, and an embarrassing one for the medical profession. The following famous quiz was given to medical doctors (which I borrowed from the excellent Deborah Bennett's *Randomness*).

A test of a disease presents a rate of 5% false positives. The disease strikes 1/1,000 of the population. People are tested at random, regardless of whether they are suspected of having the disease. A patient's test is positive. What is the probability of the patient being stricken with the disease?

Most doctors answered 95%, simply taking into account the fact that the test has a 95% accuracy rate. The answer is the conditional probability that the patient is sick and the test shows it—close to 2%. Less than one in five professionals got it right.

I will simplify the answer (using the frequency approach). Assume no false negatives. Consider that out of 1,000 patients who are administered the test, one will be expected to be afflicted with the disease. Out of a population of the remaining 999 healthy patients, the test will identify about 50 with the disease (it is 95% accurate). The correct answer should be that the probability of being afflicted with the disease for someone selected at random who presented a positive test is the following ratio:

$$\frac{\text{Number of afflicted persons}}{\text{Number of true and false positives}}$$

here 1 in 51.

Think of the number of times you will be given a medication that carries damaging side effects for a given disease you were told you had, when you may only have a 2% probability of being afflicted with it!

We Are Option Blind

As an option trader, I have noticed that people tend to undervalue options as they are usually unable to correctly mentally evaluate instruments that deliver an *uncertain* payoff, even when they are fully conscious of the mathematics. Even regulators reinforce such ignorance by explaining to people that options are a *decaying* or *wasting* asset. Options that are out of the money are deemed to *decay*, by losing their premium between two dates.

I will clarify next with a simplified (but sufficient) explanation of what an option means. Say a stock trades at $100 and that someone gives me the right (but not the obligation) to buy it at $110 one month ahead of today. This is dubbed a *call* option. It makes sense for me to *exercise* it, by asking the seller of the option to deliver me the stock at $110, only if it trades at a higher price than $110 in one month's time. If the stock goes to $120, my option will be worth $10, for I will be able to buy the stock at $110 from the option writer and sell it to the market at $120, pocketing the difference. But this does not have a very high probability. It is called *out-of-the-money*, for I have no gain from exercising it right away.

Consider that I buy the option for $1. What do I expect the value of the option to be one month from now? Most people think 0. That is not true. The option has a high probability, say 90%, of being worth 0 at expiration, but perhaps 10% probability to be worth an average of $10. Thus, selling the option to me for $1 does not provide the seller with free money. If the seller had instead bought the stock himself at $100 and waited the month, he could have sold it for $120. Making $1 now was hardly, therefore, free money. Likewise, buying it is not a wasting asset. Even professionals can be fooled. How? They confuse the expected value and the most likely scenario (here the expected value is $1 and the most likely scenario is for the option to be worth 0). They mentally overweigh the state that is the most likely, namely, that the market does not move at all. The option is simply the weighted average of the possible states the asset can take.

There is another type of satisfaction provided by the option seller. It is the steady return and the steady feeling of reward— what psychologists call *flow*. It is very pleasant to go to work in the morning with the expectation of being up some small money. It requires some strength of character to accept the expectation of bleeding a little, losing pennies on a steady basis even if the strategy is bound to be profitable over longer periods. I noticed that very few option traders can maintain what I call a "long volatility" position, namely a position that will most likely lose a small quantity of money at expiration, but is expected to make money in the long run because of occasional spurts. I discovered very few people who accepted losing $1 for most expirations and making $10 once in a while, even if the game were fair (i.e., they made the $10 more than 9.1% of the time).

I divide the community of option traders into two categories: *premium sellers* and *premium buyers*. Premium sellers (also called option sellers) sell options, and generally make steady money, like John in Chapters 1 and 5. Premium buyers do the reverse. Option sellers, it is said, eat like chickens and go to the bathroom like elephants. Alas, most option traders I encountered in my career are *premium sellers*—when they blow up it is generally other people's money.

How could professionals seemingly aware of the (simple) mathematics be put in such a position? As previously discussed, our actions are not quite guided by the parts of our brain that dictate rationality. We think with our emotions and there is no way around it. For the same reason, people who are otherwise rational engage in smoking or in fights that get them no immediate benefits; likewise people sell options even when they know that it is not a good thing to do. But things can get worse. There is a category of people, generally finance academics, who, instead of fitting their actions to their brains, fit their brains to their actions. These people go back and unwittingly cheat with the statistics to justify their actions. In my business, they fool themselves with statistical arguments to justify their option selling.

What is less unpleasant: to lose 100 times $1 or lose once $100? Clearly the second: Our sensitivity to losses decreases. So a trading policy that makes $1 a day for a long time then loses them all is actually pleasant from a hedonic standpoint, although it does not make sense economically. So there is an incentive to invent a story about the likelihood of the events and carry on such strategy.

In addition, there is the risk ignorance factor. Scientists have subjected people to tests—what I mentioned in the prologue as risk taking out of underestimating the risks rather than courage. The subjects were asked to predict a range for security prices in the future, an upper bound and a lower bound, in such a way that they would be comfortable with 98% of the security ending inside such range. Of course violations to such bound were very large, up to 30%.

Such violations arise from a far more severe problem: People overvalue their knowledge and underestimate the probability of their being wrong.

One example to illustrate further option blindness. What has more value? (a) a contract that pays you $1 million if the stock market goes down 10% on any given day in the next year; (b) a contract that pays you $1 million if the stock market goes down 10% on any given day in the next year due to a terrorist act. I expect most people to select (b).

PROBABILITIES AND THE MEDIA
(MORE JOURNALISTS)

A journalist is trained in methods to express himself rather than to plumb the depth of things—the selection process favors the most communicative, not necessarily the most knowledgeable. My medical doctor friends claim that many medical journalists do not understand anything about medicine and biology, often making mistakes of a very basic nature. I cannot confirm such statements, being myself a mere amateur (though at times a voracious reader) in medical research, but I have noticed that they almost always misunderstand the probabilities used in medical research announcements. The most common one concerns the interpretation of evidence. They most commonly get mixed up between *absence of evidence* and *evidence of absence*, a similar problem to the one we saw in Chapter 9. How? Say I test some chemotherapy, for instance Fluorouracil, for upper respiratory tract cancer, and find that it is better than a placebo, but only marginally so; that (in addition to other modalities) it improves survival from 21 per 100 to 24 per 100. Given my sample size, I may not be confident that the additional 3% survival points come from the medicine; it could be merely attributable to randomness. I would write a paper outlining my results and saying that there is no evidence of improved survival (as yet) from such medicine, and that further research would be needed. A medical journalist would pick it up and claim that one Professor N. N. Taleb found evidence that Fluorouracil *does not help*, which is entirely opposite to my intentions. Some naive doctor in Smalltown, even more uncomfortable with probabilities than the most untrained journalist, would pick it up and build a mental block against the medication, even when some researcher finally finds fresh evidence that such medicine confers a clear survival advantage.

CNBC at Lunchtime

The advent of the financial television channel CNBC presented plenty of benefits to the financial community but it also allowed a collection of extrovert practitioners long on theories to voice them in a few minutes of television time. One often sees respectable people making ludicrous (but smart-sounding) statements about properties of the stock market. Among these are statements that blatantly violate the laws of probability. One summer during which I was assiduous at the health club, I often heard statements such as "the real market is only 10% off the highs while the average stock is close to 40% off its highs," which is intended to be indicative of deep troubles or anomalies—some harbinger of bear markets.

There is no incompatibility between the fact that the average stock is down 40% from the highs while the average of all stocks (that is, the market) is down 10% from its own highs. One must consider that the stocks did not all reach their highs *at the same time*. Given that stocks are not 100% correlated, stock A might reach its maximum in January, stock B might reach its maximum in April, but the average of the two stocks A and B might reach its maximum at some time in February. Furthermore, in the event of negatively correlated stocks, if stock A is at its maximum when stock B is at its minimum, then they could both be down 40% from their maximum when the stock market is at its highs! By a law of probability called distribution of the maximum of random variables, the maximum of an average is necessarily less volatile than the average maximum.

You Should Be Dead by Now

This brings to mind another common violation of probability by prime-time TV financial experts, who may be selected for their looks, their charisma, and their presentation skills, but certainly not for their incisive minds. For instance, a fallacy that I saw commonly made by a prominent TV financial guru goes as follows: "The average American is expected to live seventy-three years. Therefore if you are sixty-eight you can expect to live five more years, and should plan accordingly." She went into precise prescriptions of how the person should invest for a five-more-years horizon. Now what if you are eighty? Is your life expectancy *minus* seven years? What these journalists confuse is the unconditional and conditional life expectancy. At birth, your unconditional life expectancy may be seventy-three years. But as you advance in age and do not die, your life expectancy increases along with your life. Why? Because other people, by dying, have taken your spot in the statistics, for expectation means average. So if you are seventy-three and are in good health, you may still have, say, nine years *in expectation*. But the expectation would change, and at eighty-two, you will have another five years, provided of course you are still alive. Even someone one hundred years old still has a positive conditional life expectation. Such a statement, when one thinks about it, is not too different from the one that says: Our operation has a mortality rate of 1%. So far we have operated on ninety-nine patients with great success; you are our one hundreth, hence you have a 100% probability of dying on the table.

TV financial planners may confuse a few people. This is quite harmless. What is far more worrying is the supply of information by nonprofessionals to professionals; it is to the journalists that we turn next.

The Bloomberg Explanations

I have, on my desk, a machine eponymously called a *Bloomberg* (after the legendary founder Michael Bloomberg). It acts as a safe e-mail service, a news service, a historical-data retrieving tool, a charting system, an invaluable analytical aid, and, not least, a screen where I can see the price of securities and currencies. I have gotten so addicted to it that I cannot operate without it, as I would otherwise feel cut off from the rest of the world. I use it to get in contact with my friends, confirm appointments, and solve some of those entertaining quarrels that put some sharpness into life. Somehow, traders who do not have a Bloomberg address do not exist for us (they have to have recourse to the more plebeian Internet). But there is one aspect of Bloomberg I would dispense with: the journalist's commentary. Why? Because they engage in explaining things and perpetuate the right-column, left-column confusion in a serious manner. Bloomberg is not the sole perpetrator; it is just that I have not been exposed to newspapers' business sections over the past decade, preferring to read real prose instead.

As I am writing these lines I see the following headlines on my Bloomberg:

→ *Dow is up 1.03 on lower interest rates.*

→ *Dollar down 0.12 yen on higher Japanese surplus.*

and so on for an entire page. If I translate it well, the journalist claims to provide an explanation for something that amounts to *perfect noise.* A move of 1.03 with the Dow at 11,000 constitutes less than a 0.01% move. Such a move does not warrant an explanation. There is nothing there that an honest person can try to explain; there are no reasons to adduce. But like apprentice professors of comparative literature, journalists being paid to provide explanations will gladly and readily provide them. The only solution is for Michael Bloomberg to stop paying his journalists for providing commentary.

Significance: How did I decide that it was perfect noise? Take a simple analogy. If you engage in a mountain bicycle race with a friend across Siberia and, a month later, beat him by one single second, you clearly cannot quite boast that you are faster than him. You might have been helped by something, or it can be just plain randomness, nothing else. That second is not in itself significant enough for someone to draw conclusions. I would not write in my pre-bedtime diary: *Cyclist A is better than cyclist B because he is fed with spinach whereas cyclist B has a diet rich in tofu. The reason I am making this inference is because he beat him by 1.3 seconds in a 3,000 mile race.* Should the difference be one week, then I could start analyzing whether tofu is the reason, or if there are other factors.

Causality: There is another problem; even assuming statistical significance, one has to accept a cause and effect, meaning that the event in the market can be linked to the cause proffered. *Post hoc ergo propter hoc* (it is the consequence because it came after). Say hospital A delivered 52% boys and hospital B delivered the same year only 48%; would you try to give the explanation that you had a boy because it was delivered in hospital A?

Causality can be very complex. It is very difficult to isolate a single cause when there are plenty around. This is called multivariate analysis. For instance, if the stock market can react to U.S. domestic interest rates, the dollar against the yen, the dollar against the European currencies, the European stock markets, the United States balance of payments, United States inflation, and another dozen prime factors, then the journalists need to look at all of these factors, look at their historical effect both in isolation and jointly, look at the stability of such influence, then, after consulting the test statistic, isolate the factor if it is possible to do so. Finally, a proper confidence level needs to be given to the factor itself; if it is less than 90% the story would be dead. I can understand why Hume was extremely obsessed with causality and could not accept such inference anywhere.

I have a trick to know if something *real* in the world is taking place. I have set up my Bloomberg monitor to display the price and percentage change of all relevant prices in the world: currencies, stocks, interest rates, and commodities. By dint of looking at the same setup for years, as I keep the currencies in the upper left corner and the various stock markets on the right, I managed to build an instinctive way of knowing if something serious is going on. The trick is to look only at the large percentage changes. Unless something moves by more than its usual daily percentage change, the event is deemed to be noise. Percentage moves are the size of the headlines. In addition, the interpretation is not linear; a 2% move is not twice as significant an event as 1%, it is rather like four to ten times. A 7% move can be several billion times more relevant than a 1% move! The headline of the Dow moving by 1.3 points on my screen today has less than one billionth of the significance of the serious 7% drop of October 1997. People might ask me: Why do I want everybody to learn some statistics? The answer is that too many people read explanations. We cannot instinctively understand the nonlinear aspect of probability.

Filtering Methods

Engineers use methods to clean up the noise from the signal in the data. Did it ever occur to you while talking to your cousin in Australia or the South Pole that the static on the telephone line could be distinguished from the voice of your correspondent? The method is to consider that when a change in amplitude is small, it is more likely to result from noise—with its likelihood of being a signal increasing exponentially as its magnitude increases. The method is called a smoothing kernel, which has been applied in Figures 11.1 and 11.2. But our auditory system is incapable of performing such a function by itself. Likewise our brain cannot see the difference between a significant price change and mere noise, particularly when it is pounded with unsmoothed journalistic noise.

We Do Not Understand Confidence Levels

Professionals forget the following reality. It is not the estimate or the forecast that matters so much as the degree of confidence with

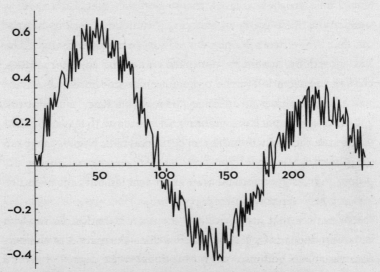

Figure 11.1   Unfiltered Data Containing Signal and Noise

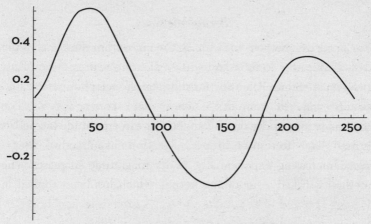

Figure 11.2   Same Data with Its Noise Removed

the opinion. Consider that you are going on a trip one fall morning and need to formulate an idea about the weather conditions prior to packing your luggage. If you expect the temperature to be 60 degrees, plus or minus 10 degrees (say in Arizona), then you would take no snow clothes and no portable electric fan. Now, what if you were going to Chicago, where you are told that the weather, while being 60 degrees, will nevertheless vary by about 30 degrees? You would have to pack winter and summer clothes. Here the expectation of the temperature carries little importance concerning the choice of clothing; it is the variance that matters. Your decision to pack is markedly different now that you are told that the variability would be around 30 degrees. Now let us push the point further; what if you were going to a planet where the expectation is also going to be around 60 degrees, but plus or minus 500 degrees? What would you pack?

We can see that my activity in the market (and other random variables) depends far less on where I think the market or the random variable is going so much as it does on the degree of error I allow around such a confidence level.

An Admission

We close this chapter with the following information: I consider myself as prone to foolishness as anyone I know, in spite of my profession and the time spent building my expertise on the subject. But here is the exception; I know that I am very, very weak on that score. My humanity will try to foil me; I have to stay on my guard. I was born to be fooled by randomness. That will be explored in Part III.

Part III

•

WAX IN MY EARS

Living with Randomitis

O dysseus, the Homerian hero, had the reputation of using guile to overcome stronger opponents. I find the most spectacular use of such guile was against no other opponent than himself.

In Book 12 of the *Odyssey*, the hero encounters the sirens, on an island not far from the rocks of Charybdis and Scylla. Their songs are known to charm the sailors into madness, causing them irresistibly to cast themselves into the sea off the sirens' coast, and perish. The indescribable beauty of the sirens' songs is contrasted with the moldering corpses of sailors who strayed into the area around them. Odysseus, forewarned by Circe, contrives the following ruse. He fills the ears of all his men with wax, to the point of total deafness, and has himself tied to the mast. The sailors are under strict instructions not to release him. As they approach the sirens' island, the sea is calm and over the water comes the sound of a music so ravishing that Odysseus struggles to get loose, expending an inordinate amount of energy to unrestrain himself. His men tie him even further, until they are safely past the poisoned sounds.

The first lesson I took from the story is not to even attempt to be Odysseus. He is a mythological character and I am not. He can be tied to the mast; I can merely reach the rank of a sailor who needs to have his ears filled with wax.

I AM NOT SO INTELLIGENT

The epiphany I had in my career in randomness came when I understood that I was not intelligent enough, nor strong enough, to even try to fight my emotions. Besides, I believe that I need my emotions to formulate my ideas and get the energy to execute them.

I am just intelligent enough to understand that I have a predisposition to be fooled by randomness—and to accept the fact that I am rather emotional. I am dominated by my emotions—but as an aesthete, I am happy about that fact. I am just like every single character whom I ridiculed in this book. Not only that, but I may be even worse than them because there may be a negative correlation between beliefs and behavior (recall Popper the man). The difference between me and those I ridicule is that I try to be aware of it. No matter how long I study and try to understand probability, my emotions will respond to a different set of calculations, those that my unintelligent genes want me to handle. If my brain can tell the difference between noise and signal, my heart cannot.

Such unintelligent behavior does not just cover probability and randomness. I do not think I am reasonable enough to avoid getting angry when a discourteous driver blows his horn at me for being one nanosecond late after a traffic light turns green. I am fully aware that such anger is self-destructive and offers no benefit, and that if I were to develop anger for every idiot around me doing something of the sort, I would be long dead. These small daily emotions are not rational. But we need them to function properly. We are designed to respond to hostility with hostility. I have enough enemies to add some spice to my life, but I sometimes wish I had a few more (I rarely go to the movies and need the entertainment). Life would be unbearably bland if we had no enemies on whom to waste efforts and energy.

The good news is that there are tricks. One such trick is to avoid eye contact (through the rearview mirror) with other persons in such traffic encounters. Why? Because when you gaze into someone's eyes, a different part of your brain, the more emotional one, is activated and engaged as the result of the interaction. I try to imagine that the other person is a Martian, rather than a human being. It works sometimes—but it works best when the person presents the appearance of being from a different species. How? I am an avid road cyclist. Recently, as I was riding along with other cyclists, slowing down traffic in a rural area, a small woman in a giant sports utility vehicle opened her window and heaped curses at us. Not only did it not upset me but I did not even interrupt my thoughts to pay attention. When I am on my bicycle, people in large trucks become a variety of dangerous animals, capable of threatening me but incapable of making me angry.

I have, like anyone with strong opinions, a collection of critics among finance academics and economists, annoyed by my attacks on their misuse of probability and unhappy about my branding them as pseudoscientists. I am incapable of taming my emotions when reading their comments. The best I can do is just not read them. Likewise with journalists. Not reading their discussions of markets spares me plenty of emotional expenditure. I will do the same with unsolicited comments on this book. Wax in my ears.

WITTGENSTEIN'S RULER

What is the mechanism that should convince authors to avoid reading comments on their work, except for those they solicit from specified persons for whom they have intellectual respect? The mechanism is a probabilistic method called conditional information: Unless the source of the statement has extremely high qualifications, the statement will be more revealing of the author than the information intended by him. This applies, of course, to matters of judgment. A book review, good or bad, can be far more descriptive of the reviewer than informational about the book it-self. This mechanism I also call Wittgenstein's ruler: *Unless you have confidence in the ruler's reliability, if you use a ruler to measure a table you may also be using the table to measure the ruler.* The less you trust the ruler's reliability (in probability called the *prior*), the more information you are getting about the ruler and the less about the table. The point extends way beyond information and probability. This conditionality of information is central in episte-mology, probability, even in studies of consciousness. We will see later extensions with "ten sigma" problems.

The point carries practical implications: The information from an anonymous reader on Amazon.com is all about the person, while that of a qualified person, is going to be all about the book. This plays equally in court: Take the O. J. Simpson trial once again. One of the jurors said, "There was not enough blood," meaning to assess the statistical evidence of what was offered: Such statement reveals very little about the statistical evidence as compared with what it shows about the author of the statement's ability to make a valid inference. Had the juror been a forensic expert, the ratio of information would have tilted the other way.

The problem is that while such reasoning is central to my think-ing, my brain knows it though not my heart: My emotional system does not understand Wittgenstein's ruler. I can offer the following evidence: A compliment is always pleasant, regardless of its au-thorship—something manipulators know rather well. Likewise with book reviews or comments on my risk-management strategy.

THE ODYSSEAN MUTE COMMAND

Recall that the accomplishment from which I derive the most pride is my weaning myself from television and the news media. I am currently so weaned that it actually costs me more energy to watch television than to perform any other activity, like, say, writ-

ing this book. But this did not come without tricks. Without tricks
I would not escape the toxicity of the information age. In the trading room of my company, I have the television set turned on all
day with the financial news channel CNBC staging commentator
after commentator and CEO after CEO murdering rigor all day
long. What is the trick? I have the volume turned completely off.
Why? Because when the television set is silent, the babbling person looks ridiculous, exactly the opposite effect as when the sound
is on. One sees a person with moving lips and contortions in his
facial muscles, taking themselves seriously—but no sound comes
out. We are visually but not auditorily intimidated, which causes a
dissonance. The speaker's face expresses some excitement, but
since no sound comes out, the exact opposite is conveyed. This is
the sort of contrast the philosopher Henri Bergson had in mind in
his *Treatise on Laughter*, with his famous description of the gap between the seriousness of a gentleman about to walk on a banana
skin and the comical aspect of the situation. Television pundits
lose their intimidating effect; they even look ridiculous. They seem
to be excited about something terribly unimportant. Suddenly
pundits become clowns, which is a reason the writer Graham
Greene refused to go on television.

I had this idea of stripping people of language while, on a trip, I
listened (while brutally jet-lagged) to a speech in Cantonese, a language I do not understand, without the benefit of translation.
Since I had no possible clue about his subject, the animated orator
lost a large share of his dignity. The idea came to me that perhaps
I could use a built-in bias, here prejudice, to offset another built-
in bias, our predisposition to take information seriously. It seems
to work.

This part, the conclusion of this book, presents the human aspect of dealing with uncertainty. I have personally failed in achiev-
ing a general insulation from randomness, but I have managed a
few tricks.

•

Twelve

GAMBLERS' TICKS AND PIGEONS IN A BOX

On gamblers' ticks crowding up my life. Why bad taxi-cab English can help you make money. How I am the fool of all fools, except that I am aware of it. Dealing with my genetic unfitness. No boxes of chocolate under my trading desk.

TAXI-CAB ENGLISH AND CAUSALITY

First, a flashback in time to my early days as a trader in New York. Early in my career, I worked at Credit Suisse First Boston, then located in the middle of the block between Fifty-second and Fifty-third streets, between Madison and Park Avenue. It was called a Wall Street firm, in spite of its Midtown location—I used to claim to work "on Wall Street" in spite of having been lucky enough to set foot only twice on the physical Wall Street, one of the most repulsive areas I have visited east of Newark, New Jersey.

Then, in my twenties, I lived in a book-choked (but otherwise rather bare) apartment on Manhattan's Upper East Side. The bareness was not ideological; it was simply because I never managed to enter a furniture store, as I would eventually stop at a bookstore along the way and haul bags of books instead. As can be expected, the kitchen was devoid of any form of food and utensils, save for a defective espresso machine, as I learned to cook only very recently (even then . . .).

I went to work every morning in a yellow cab, which dropped me off at the corner of Park Avenue and Fifty-third Street. Cab drivers in New York City are known to be rather untamed and universally unfamiliar with the geography of the place, but, on occasion, one can find a cab driver who is both unacquainted with the city and skeptical of the universality of the laws of arithmetic. One day I had the misfortune (or perhaps the fortune, as we will see) to ride with a driver who did not seem capable of handling any language known to me, which includes taxi-cab English. I tried to help him navigate south between Seventy-fourth Street and Fifty-third Street, but he stubbornly continued the journey an additional block south, forcing me to use the Fifty-second Street entrance. That day, my trading portfolio made considerable profits, owing to considerable turmoil in currencies; it was then the best day of my young career.

The next day, as usual, I hailed a cab from the corner of Seventy-fourth Street and Third Avenue. The previous driver was nowhere in sight, perhaps deported back to the old country. Too bad; I was gripped with the unexplainable desire to pay him back for the favor he had done me and surprise him with a gigantic tip. Then I caught myself instructing the new cab driver to take me to the northeast corner of Fifty-second Street and Park Avenue, exactly where I was dropped off the day before. I was taken aback by my own words . . . but it was too late.

When I looked at my reflection in the elevator's mirror, it dawned on me that I wore the exact same tie as the day before—

with the coffee stains from the previous day's fracas (my only addiction is to coffee). There was someone in me who visibly believed in a strong causal link between my use of the entrance, my choice of tie, and the previous day's market behavior. I was disturbed for acting like a fake, like an actor playing some role that was not his. I felt that I was an impostor. On the one hand, I talked like someone with strong scientific standards, a probabilist focused on his craft. On the other, I had closed superstitions just like one of these blue-collar pit traders. Would I have to go buy a horoscope next?

A little brooding revealed that my life until then had been governed by mild superstitions, me the expert in options, the dispassionate calculator of probabilities, a rational trader! It was not the first time that I had acted on mild superstitions of a harmless nature, which I believed were instilled in me by my Eastern Mediterranean roots: One does not grab the salt shaker from the hand of another person risking a falling out; one is to knock on wood upon receiving a compliment; plus many other Levantine beliefs passed on for a few dozen centuries. But like many things that brew and spread around the ancient pond, these beliefs I had taken with a fluctuating mixture of solemnity and mistrust. We consider them more like rituals than truly important actions meant to stave off undesirable turns of the goddess Fortuna—superstitions can instill some poetry in daily life.

The worrying part was that it was the first time that I noticed superstitions creeping into my professional life. My profession is to act like an insurance company, stringently computing the odds based on well-defined methods, taking advantage of other people when they are less rigorous, get blinded by some "analysis," or act with the belief that they are chosen by destiny. But there was too much randomness flooding my occupation.

I detected the rapid accumulation of what are called "gamblers' ticks" surreptitiously developing in my behavior—though minute and barely detectable. Until then these small ticks had escaped me. My mind seemed to be constantly trying to detect a statistical

connection between some of my facial expressions and the outcome of events. For example, my income started to increase after
I discovered my slight nearsightedness and started wearing glasses.
Although glasses were not quite necessary, nor even useful, except
for night driving, I kept them on my nose as I unconsciously acted
as if I believed in the association between performance and
glasses. To my brain such statistical association was as spurious as
it can get, owing to the reduced sample size (here a single instance), yet this native statistical instinct did not seem to benefit
from my expertise in hypothesis testing.

Gamblers are known to develop some behavioral distortions as
a result of some pathological association between a betting outcome and some physical move. "Gambler" is about the most
derogatory term that could be used in my derivatives profession.
As an aside, gambling to me is best defined as an activity where the
agent gets a thrill when confronting a random outcome, regardless
of whether he has the odds stacked in his favor or against him.
Even when the odds are clearly stacked against the gambler, he
sometimes transcends the odds by believing that destiny selected
him in some manner. This shows in the very sophisticated people
one meets in casinos where they normally should not be found. I
even ran into world-class probability experts who had a gambling
habit on the side, throwing all of their knowledge to the wind. For
example, a former colleague of mine, one of the most intelligent
people I have ever met, frequently went to Las Vegas, and seemed
to be such a turkey that the casino provided him with compli-
mentary luxury suites and transportation. He even consulted a
fortune teller prior to taking large trading positions and tried to
get reimbursed by our employer.

THE SKINNER PIGEON EXPERIMENT

At twenty-five, I was totally ignorant of the behavioral sciences. I
had been fooled by my education and culture into believing that

my superstitions were cultural, and that, consequently, they could be shed through the exercise of so-called reason. Taken at the general level of society, modern life would eliminate them as science and logic would enter. But in my case, I was over time getting more sophisticated intellectually, but the floodgates of randomness were bursting and I was becoming more superstitious.

These superstitions needed to be biological—but I was brought up in an era when the dogma was that it was nurture, rarely nature, that was the culprit. Clearly, there was nothing cultural about my link between my wearing glasses and a random market outcome. There was nothing cultural in my link between my use of entrance and my performance as a trader. There was nothing cultural in my wearing the same tie as the day before. Something in us has not developed properly over the past thousand years and I was dealing with the remnant of our old brain.

To probe the point further, we need to look at such formations of causal associations in the lower forms of life. The famous Harvard psychologist B. F. Skinner constructed a box for rats and pigeons, equipped with a switch that the pigeon can operate by pecking. In addition, an electrical mechanism delivers food into the box. Skinner designed the box in order to study more general properties of the behavior of a collection of nonhumans, but it was in 1948 that he had the brilliant idea of ignoring the lever and focusing on the food delivery. He programmed it to deliver food at random to the famished birds.

He saw quite astonishing behavior on the part of the birds; they developed an extremely sophisticated rain-dance type of behavior in response to their ingrained statistical machinery. One bird swung its head rhythmically against a specific corner of the box, others spun their heads counterclockwise; literally all of the birds developed a specific ritual that progressively became hardwired into their mind as linked to their feeding.

This problem has a more worrying extension; we are not made

to view things as independent from each other. When viewing two events A and B, it is hard not to assume that A causes B, B causes A, or both cause each other. Our bias is immediately to establish a causal link. While to a budding trader this results in hardly any worse costs than a few pennies in cab fare, it can draw the scientist into spurious inference. For it is harder to act as if one were ignorant than as if one were smart; scientists know that it is emotionally harder to reject a hypothesis than to accept it (what are called type I and type II errors)—quite a difficult matter when we have such sayings as *felix qui po¨tuit cognoscere causas* (happy is he who understands what is behind things). It is very hard for us to just shut up. We are not cut out for it. Popper or not, we take things too seriously.

PHILOSTRATUS REDUX

I offered no solution to the problem of statistical inference at a low resolution. I discussed in Chapter 3 the technical difference between noise and meaning—but it is time to discuss the execution. The Greek philosopher Pyrrho, who advocated a life of equanimity and indifference, was criticized for failing to keep his composure during a critical circumstance (he was chased by an ox). His answer was that he found it sometimes difficult to rid himself of his humanity. If Pyrrho cannot stop being human, I do not see why the rest of us should resemble the rational man who acts perfectly under uncertainty as propounded by economic theory. I discovered that much of the rationally obtained results using my computations of the various probabilities do not register deeply enough to impact my own conduct. In other words, I acted like the doctor in Chapter 11 who knew of the 2% probability of the disease, but somehow unwittingly treated the patient as if the ailment had a 95% probability of being there. My brain and my instinct were not acting in concert.

The details are as follows. As a rational trader (all traders boast so) I believe, as I discussed before, that there is a difference between noise and signal, and that noise needs to be ignored while a signal needs to be taken seriously. I use elementary (but robust) methods that allow me to calculate the expected noise and signal composition of any fluctuation in my trading performance. For example, after registering a profit of $100,000 on a given strategy, I may assign a 2% probability to the hypothesis of the strategy being profitable and 98% probability to the hypothesis that the performance may be the result of mere noise. A gain of $1 million on the other hand, certifies that the strategy is a profitable one, with a 99% probability. A rational person would act accordingly in the selection of strategies, and set his emotions in accordance with his results. Yet I have experienced leaps of joy over results that I knew were mere noise, and bouts of unhappiness over results that did not carry the slightest degree of statistical significance. I cannot help it, but I am emotional and derive most of my energy from my emotions. So the solution does not reside in taming my heart.

Since my heart does not seem to agree with my brain, I need to take serious action to avoid making irrational trading decisions, namely, by denying myself access to my performance report unless it hits a predetermined threshold. This is no different from the divorce between my brain and my appetite when it comes to the consumption of chocolate. I generally deal with it by ascertaining that there are no chocolate boxes under my trading desk.

One of the most irritating conversations I've had is with people who lecture me on *how I should* behave. Most of us know pretty much *how we should* behave. It is the execution that is the problem, not the absence of knowledge. I am tired of the moralizing slow-thinkers who pound me with platitudes like I should floss daily, eat my regular apple, and visit the gym outside of the New Year's resolution. In the markets the recommendation would be to ignore the noise component in the performance. We need tricks to get us there but before that we need to accept the fact that we are mere animals in need of lower forms of tricks, not lectures.

Finally, I consider myself lucky for not having a cigarette addiction. For the best way to understand how we could be rational in our perception of the risks and probabilities and, at the same time, be foolish while acting on them would be to have a conversation with a cigarette smoker. For few cigarette smokers remain unaware of the lung cancer rates in their population. If you remain unconvinced, take a look at the huddling smoking crowd outside the service entrance of the Memorial Sloan-Kettering Cancer Center in New York City's Upper East Side. You will see dozens of cancer nurses (and, perhaps, doctors) standing outside the entrance with a cigarette in hand as hopeless patients are wheeled in for their treatments.

•

Thirteen

CARNEADES COMES TO ROME:
ON PROBABILITY AND SKEPTICISM

Cato the censor sends Carneades packing. Monsieur de Norpois does not remember his old opinions. Beware the scientist. Marrying ideas. The same Robert Merton putting the author on the map. Science evolves from funeral to funeral.

A sk your local mathematician to define probability; he would most probably show you how to compute it. As we saw in Chapter 3 on probabilistic introspection, probability is not about the odds, but about the belief in the existence of an alternative outcome, cause, or motive. Recall that mathematics is a tool to meditate, not compute. Again, let us go back to the elders for more guidance—for probabilities were always considered by them as nothing beyond a subjective, and fluid, measure of beliefs.

CARNEADES COMES TO ROME

Around 155 B.C., the Greek postclassical philosopher Carneades of Cyrene came to Rome as one of the three Athenian ambassadors who came to beg the Roman Senate for a political favor. A fine had been levied against the citizens of their city, and they wanted to convince Rome that it was unfair. Carneades represented the Academy, the same argumentative open-air institution where three centuries before, Socrates drove his interlocutors to murder him just to get some respite from his arguments. It was now called the New Academy, was no less argumentative, and had the reputation of being the hotbed of skepticism in the ancient world.

On the much anticipated day of his oration, he stood up and delivered a brilliantly argued harangue in praise of justice and how devolving it should be at the top of our motives. The Roman audience was spellbound. It was not just his charisma; the audience was swayed by the strength of the arguments, the eloquence of the thought, the purity of the language, and the energy of the speaker. But that was not the point he wanted to drill.

The next day, Carneades came back, stood up, and established the doctrine of uncertainty of knowledge in the most possibly convincing way. How? By proceeding to contradict and refute with no less swaying arguments what he had established so convincingly the day before. He managed to persuade the very same audience and in the same spot that justice should be way down on the list of motivations for human undertakings.

Now the bad news. Cato the elder (the "censor") was among the audience, already quite old, and no more tolerant than he had been during his office of censor. Enraged, he persuaded the Senate to send the three ambassadors packing lest their argumentative spirit muddle the spirit of the youth of the Republic and weaken the military culture. (Cato had banned during his office of censorship all Greek rhetoricians from establishing residence in Rome. He was too much a no-nonsense type of person to accept their introspective expansions.)

Carneades was not the first skeptic in classical times, nor was he the first to teach us the true notion of probability. But this incident remains the most spectacular in its impact on generations of rhetoricians and thinkers. Carneades was not merely a skeptic; he was a dialectician, someone who never committed himself to any of the premises from which he argued, or to any of the conclusions he drew from them. He stood all his life against arrogant dogma and belief in one sole truth. Few credible thinkers rival Carneades in their rigorous skepticism (a class that would include the medieval Arab philosopher Al Gazali, Hume, and Kant—but only Popper came to elevate his skepticism to an all-encompassing scientific methodology). As the skeptics' main teaching was that nothing could be accepted with certainty, conclusions of various degrees of probability could be formed, and these supplied a guide to conduct.

Stepping further back in time and searching for the first known uses of probabilistic thinking in history, we find it harks back to sixth-century (B.C.) Greek Sicily. There, the notion of probability was used in a legal framework by the very first rhetoricians who, when arguing a case, needed to show the existence of a doubt concerning the certainty of the accusation. The first known rhetorician was a Syracusean named Korax, who engaged in teaching people how to argue from probability. At the core of his method was the notion of the *most probable*. For example, the ownership of a piece of land, in the absence of further information and physical evidence, should go to the person after whose name it is best known. One of his indirect students, Gorgias, took this method of argumentation to Athens, where it flourished. It is the establishment of such *most probable* notions that taught us to view the possible contingencies as distinct and separable events with probabilities attached to each one of them.

Probability, the Child of Skepticism

Until the Mediterranean basin was dominated with monotheism, which led to the belief in some form of uniqueness of the truth (to be superceded later by episodes of communism), skepticism had gained currency among many major thinkers—and certainly permeated the world. The Romans did not have a religion *per se;* they were too tolerant to accept a given truth. Theirs was a collection of a variety of flexible and syncretic superstitions. I will not get too theological, except to say that we had to wait for a dozen centuries in the Western world to espouse critical thinking again. Indeed, for some strange reason during the Middle Ages, Arabs were critical thinkers (through their postclassical philosophical tradition) when Christian thought was dogmatic; then, after the Renaissance, the roles mysteriously reversed.

One author from antiquity who provides us evidence of such thinking is the garrulous Cicero. He preferred to be guided by probability than allege with certainty—very handy, some said, because it allowed him to contradict himself. This may be a reason for us, who have learned from Popper how to remain self-critical, to respect him more, as he did not hew stubbornly to an opinion for the mere fact that he had voiced it in the past. Indeed your average literature professor would fault him for his contradictions and his change of mind.

It was not until modern times that such desire to be free from our own past statements emerged. Nowhere was it made more eloquently than in rioting student graffiti in Paris. The student movement that took place in France in 1968, with the youth no doubt choking under the weight of years of having to sound intelligent and coherent, produced, among other jewels, the following demand:

We demand the right to contradict ourselves!

MONSIEUR DE NORPOIS' OPINIONS

Modern times provide us with a depressing story. Self-contradiction is made culturally to be shameful, a matter that can prove disastrous in science. Marcel Proust's novel *In Search of Time Lost* features a semiretired diplomat, Marquis de Norpois, who, like all diplomats before the advent of the fax machine, was a socialite who spent considerable time in salons. The narrator of the novel sees Monsieur de Norpois openly contradicting himself on some issue (some prewar rapprochement between France and Germany). When reminded of his previous position, Monsieur de Norpois did not seem to recall it. Proust reviles him:

> Monsieur de Norpois was not lying. He had just forgotten. One forgets rather quickly what one has not thought about with depth, what has been dictated to you by imitation, by the passions surrounding you. These change, and with them so do your memories. Even more than diplomats, politicians do not remember opinions they had at some point in their lives and their fibbings are more attributable to an excess of ambition than a lack of memory.

Monsieur de Norpois is made to be ashamed of the fact that he expressed a different opinion. Proust did not consider that the diplomat might have changed his mind. We are supposed to be faithful to our opinions. One becomes a traitor otherwise.

Now I hold that Monsieur de Norpois should be a trader. One of the best traders I have ever encountered in my life, Nigel Babbage, has the remarkable attribute of being completely free of any path dependence in his beliefs. He exhibits absolutely no embarrassment buying a given currency on a pure impulse, when only hours ago he might have voiced a strong opinion as to its future weakness. What changed his mind? He does not feel obligated to explain it.

The public person most visibly endowed with such a trait is George Soros. One of his strengths is that he revises his opinion rather rapidly, without the slightest embarrassment. The following anecdote illustrates Soros' ability to reverse his opinion in a flash. The French playboy trader Jean-Manuel Rozan discusses the following episode in his autobiography (disguised as a novel in order to avoid legal bills). The protagonist (Rozan) used to play tennis in the Hamptons on Long Island with Georgi Saulos, an "older man with a funny accent," and sometimes engage in discussions about the market, not initially knowing how important and influential Saulos truly was. One weekend, Saulos exhibited in his discussion a large amount of bearishness, with a complicated series of arguments that the narrator could not follow. He was obviously short the market. A few days later, the market rallied violently, making record highs. The protagonist worried about Saulos, and asked him at their subsequent tennis encounter if he was hurt. "We made a killing," Saulos said. "I changed my mind. We covered and went very long."

It was this very trait that, a few years later, affected Rozan negatively and almost cost him a career. Soros gave Rozan in the late 1980s $20 million to speculate with (a sizeable amount at the time), which allowed him to start a trading company (I was almost dragged into it). A few days later, as Soros was visiting Paris, they discussed markets over lunch. Rozan saw Soros becoming distant. He then completely pulled the money, offering no explanation. What characterizes real speculators like Soros from the rest is that their activities are devoid of path dependence. They are totally free from their past actions. Every day is a clean slate.

Path Dependence of Beliefs

There is a simple test to define path dependence of beliefs (economists have a manifestation of it called the endowment effect).

Say you own a painting you bought for $20,000, and owing to rosy conditions in the art market, it is now worth $40,000. If you owned no painting, would you still acquire it at the current price? If you would not, then you are said to be married to your position. There is no rational reason to keep a painting you would not buy at its current market rate—only an emotional investment. Many people get married to their ideas all the way to the grave. Beliefs are said to be path dependent if the sequence of ideas is such that the first one dominates.

There are reasons to believe that, for evolutionary purposes, we may be programmed to build a loyalty to ideas in which we have invested time. Think about the consequences of being a good trader outside of the market activity, and deciding every morning at 8 a.m. whether to keep the spouse or part with him or her for a better emotional investment elsewhere. Or think of a politician who is so rational that, during a campaign, he changes his mind on a given matter because of fresh evidence and abruptly switches political parties. That would make rational investors who evaluate trades in a proper way a genetic oddity—perhaps a rare mutation. Researchers found that purely rational behavior on the part of humans can come from a defect in the amygdala that blocks the emotions of attachment, meaning that the subject is, literally, a psychopath. Could Soros have a genetic flaw that makes him rational as a decision maker?

Such trait of absence of marriage to ideas is indeed rare among humans. Just as we do with children, we support those in whom we have a heavy investment of food and time until they are able to propagate our genes, so we do with ideas. An academic who became famous for espousing an opinion is not going to voice anything that can possibly devalue his own past work and kill years of investment. People who switch parties become traitors, renegades, or, worst of all, apostates (those who abandoned their religion were punishable by death).

COMPUTING INSTEAD OF THINKING

There is another story of probability other than the one I introduced with Carneades and Cicero. Probability entered mathematics with gambling theory, and stayed there as a mere computational device. Recently, an entire industry of "risk measurers" emerged, specializing in the application of these probability methods to assess risks in the social sciences. Certainly, the odds in games where the rules are clearly and explicitly defined are computable and the risks consequently measured. But not in the real world. For mother nature did not endow us with clear rules. The game is not a deck of cards (we do not even know how many colors there are). But somehow people "measure" risks, particularly if they are paid for it. I have already discussed Hume's problem of induction and the occurrence of black swans. Here I introduce the scientific perpetrators.

Recall that I have waged a war against the charlatanism of some prominent financial economists for a long time. The points are as follows. One Harry Markowitz received something called the Nobel Memorial Prize in Economics (which in fact is not even a Nobel Prize, as it is granted by the Swedish Central Bank in honor of Alfred Nobel—it was never in the will of the famous man). What is his achievement? Creating an elaborate method of computing *future* risk if one knows *future* uncertainty; in other words, if the world had clearly defined rules one picks up in a rulebook of the kind one finds in a Monopoly package. Now, I explained the point to a cab driver who laughed at the fact that someone ever thought that there was any scientific method to understanding markets and predicting their attributes. Somehow when one gets involved in financial economics, owing to the culture of the field, one becomes likely to forget these basic facts (pressure to publish to keep one's standing among the other academics).

An immediate result of Dr. Markowitz's theory was the near collapse of the financial system in the summer of 1998 (as we saw in

Chapters 1 and 5) by Long Term Capital Management ("LTCM"), a Greenwich, Connecticut, fund that had for principals two of Dr. Markowitz's colleagues, "Nobels" as well. They are Drs. Robert Merton (the one in Chapter 3 trouncing Shiller) and Myron Scholes. Somehow they thought they could scientifically "measure" their risks. They made absolutely no allowance in the LTCM episode for the possibility of their not understanding markets and their methods being wrong. That was not a hypothesis to be considered. I happen to specialize in black swans. Suddenly I started getting some irritating fawning respect. Drs. Merton and Scholes helped put your humble author on the map and caused interest in his ideas. The fact that these "scientists" pronounced the catastrophic losses a "ten sigma" event reveals a Wittgenstein's ruler problem: Someone saying this is a ten sigma either (a) knows what he is talking about with near perfection (the prior assumption is that it has one possibility of being unqualified in several billion billions), knows his probabilities, and it is an event that happens once every several times the history of the universe; or (b) just does not know what he is talking about when he talks about probability (with a high degree of certainty), and it is an event that has a probability higher than once every several times the history of the universe. I will let the reader pick from these two mutually exclusive interpretations which one is more plausible.

Note that the conclusions also reflect on the Nobel committee who sanctified the ideas of the gentlemen involved: Conditional on these events, did they make a mistake or were these events unusual? Is the Nobel committee composed of infallible judges? Where is Charles Sanders Peirce to talk to us about papal infallibilism? Where is Karl Popper to warn us against taking science—and scientific institutions—seriously? In a few decades will we look upon the Nobel economics committee with the same smirk as when we look at the respected "scientific" establishments of the Middle Ages that promoted (against all observational evidence) the idea that the heart was a center of heat? We have been getting things wrong in the past and we laugh at our past institutions; it is time to figure out that we should avoid enshrining the present ones.

One would think that when scientists make a mistake, they develop a new science that incorporates what has been learned from it. When academics blow up trading, one would expect them to integrate such information in their theories and make some heroic statement to the effect that they were wrong, but that now they have learned something about the real world. Nothing of the sort. Instead they complain about the behavior of their counterparts in the market who pounced on them like vultures, thus exacerbating their downfall. Accepting what has happened, clearly the courageous thing to do, would invalidate the ideas they have built throughout an entire academic career. All of the principals who engaged in a discussion of the LTCM events partook of a masquerade of science by adducing *ad hoc* explanations and putting the blame on a rare event (problem of induction: How did they know it was a rare event?). They spent their energy defending themselves rather than trying to make a buck with what they learned. Again, compare them with Soros, who walks around telling whoever has the patience to listen to him that he is fallible. My lesson from Soros is to start every meeting at my boutique by convincing everyone that we are a bunch of idiots who know nothing and are mistake-prone, but happen to be endowed with the rare privilege of knowing it.

The scientist's behavior while facing the refutation of his ideas has been studied in depth as part of the so-called attribution bias. You attribute your successes to skills, but your failures to randomness. This explains why these scientists attributed their failures to the "ten sigma" rare event, indicative of the thought that they were right but that luck played against them. Why? It is a human heuristic that makes us actually believe so in order not to kill our self-esteem and keep us going against adversity.

We have known about this wedge between performance and self-assessment since 1954, with Meehl's study of experts comparing their perceived abilities to their statistical ones. It shows a substantial discrepancy between the objective record of people's success in prediction tasks and the sincere beliefs of these people about the quality of their performance. The attribution bias has another effect: It gives people the illusion of being better at what they do, which explains the findings that 80 to 90% of people think that they are above the average (and the median) in many things.

FROM FUNERAL TO FUNERAL

I conclude with the following saddening remark about scientists in the soft sciences. People confuse science and scientists. Science is great, but individual scientists are dangerous. They are human; they are marred by the biases humans have. Perhaps even more. For most scientists are hard-headed, otherwise they would not derive the patience and energy to perform the Herculean tasks asked of them, like spending eighteen hours a day perfecting their doctoral thesis.

A scientist may be forced to act like a cheap defense lawyer rather than a pure seeker of the truth. A doctoral thesis is "defended" by the applicant; it would be a rare situation to see the student change his mind upon being supplied with a convincing argument. But science is better than scientists. It was said that science evolves from funeral to funeral. After the LTCM collapse, a new financial economist will emerge, who will integrate such knowledge into his science. He will be resisted by the older ones, but, again, they will be much closer to their funeral date than he.

•

Fourteen

BACCHUS ABANDONS ANTONY

Montherlant's death. Stoicism is not the stiff upper lip, but the illusion of victory of man against randomness. It is so easy to be heroic. Randomness and personal elegance.

When the classicist aristocratic French writer Henry de Montherlant was told that he was about to lose his eyesight to a degenerative disease, he found it most appropriate to take his own life. Such is the end that becomes a classicist. Why? Because the stoic's prescription was precisely to elect what one can do to control one's destiny in front of a random outcome. At the end, one is allowed to choose between no life at all and what one is given by destiny; we always have an option against uncertainty. But such an attitude is not limited to stoics; both competing sects in the ancient world, stoicism and

Epicureanism, recommended such control (the difference between the two resides in minor technicalities—neither philosophies meant then what is commonly accepted today in middlebrow culture).

Being a hero does not necessarily mean such an extreme act as getting killed in battle or taking one's life—the latter is only recommended in a narrow set of circumstances and considered cowardly otherwise. Having control over randomness can be expressed in the manner in which one acts in the small and the large. Recall that epic heroes were judged by their actions, not by the results. No matter how sophisticated our choices, how good we are at dominating the odds, randomness will have the last word. We are left only with dignity as a solution—dignity defined as the execution of a protocol of behavior that does not depend on the immediate circumstance. It may not be the optimal one, but it certainly is the one that makes us feel best. *Grace under pressure*, for example. Or in deciding not to toady up to someone, whatever the reward. Or in fighting a duel to save face. Or in signaling to a prospective mate during courtship: "Listen, I have a crush on you; I am obsessed with you, but I will not do a thing to compromise my dignity. Accordingly, the slightest snub and you will never see me again."

This last chapter will discuss randomness from a totally new angle; philosophical but not the *hard* philosophy of science and epistemology as we saw in Part I with the *black swan problem*. It is a more archaic, *softer* type of philosophy, the various guidelines that the ancients had concerning the manner in which a man of virtue and dignity deals with randomness—there was no real *religion* at the time (in the modern sense). It is worthy of note that before the spread of what can be best called Mediterranean monotheism, the ancients did not believe enough in their prayers to influence the course of destiny. Their world was dangerous, fraught with invasions and reversals of fortune. They needed substantial prescriptions in dealing with randomness. It is such beliefs that we will outline next.

NOTES ON JACKIE O.'S FUNERAL

If a stoic were to visit us, he would feel represented by the following poem. To many (sophisticated) lovers of poetry, one of the greatest poets who ever breathed is C. P. Cavafy. Cavafy was an Alexandrian Greek civil servant with a Turkish or Arabic last name who wrote almost a century ago in a combination of classical and modern Greek a lean poetry that seems to have eluded the last fifteen centuries of Western literature. Greeks treasure him like their national monument. Most of his poems take place in Syria (his Grecosyrian poems initially drew me to him), Asia Minor, and Alexandria. Many people believe it worth learning formal semiclassical Greek just to savor his poems. Somehow their acute aestheticism stripped of sentimentality provides a relief from centuries of mawkishness in poetry and drama. He provides a classical relief for those of us who were subjected to the middle-class-valued melodrama as represented by Dickens's novels, romantic poetry, and Verdi's operas.

I was surprised to hear that Maurice Tempelsman, last consort of Jackie Kennedy Onassis, read Cavafy's valedictory "Apoleipein o Theos Antonion" ("The God Abandons Antony") at her funeral. The poem addresses Marc Antony, who has just lost the battle against Octavius and was forsaken by Bacchus, the god who until then had protected him. It is one of the most elevating poems I have ever read, beautiful because it was the epitome of such dignified aestheticism—and because of the gentle but edifying tone of the voice of the narrator advising a man who had just received a crushing reversal of fortune.

The poem addresses Antony, now defeated and betrayed (according to legend, even his horse deserted him to go to his enemy

Octavius). It asks him to just bid her farewell, Alexandria the city that is leaving him. It tells him not to mourn his luck, not to enter denial, not to believe that his ears and eyes are deceiving him. Antony, do not degrade yourself with empty hopes. Antony,

Just listen while shaken by emotion but not with the coward's imploration and complaints.

While shaken with emotion. No stiff upper lip. There is nothing wrong and undignified with emotions—we are cut to have them. What is wrong is not following the heroic or, at least, the dignified path. That is what *stoicism* truly means. It is the attempt by man to get even with probability. I need not be nasty at all and break the spell of the poem and its message, but I cannot resist some cynicism. A couple of decades later, Cavafy, while dying of throat cancer, did not quite follow the prescription. Deprived of his voice by the surgeons, he used to randomly enter undignified spells of crying and cling to his visitors, preventing them from leaving his death room.

Some history. I said that stoicism has rather little to do with the stiff-upper-lip notion that we believe it means. Started as an intellectual movement in antiquity by a Phoenician Cypriot, Zeno of Kition, it developed by Roman time into a life based on a system of virtues—in the ancient sense when virtue meant *virtu*, the sort of belief in which virtue is its own reward. There developed a social model for a stoic person, like the gentlemen in Victorian England. Its tenets can be summarized as follows: The stoic is a person who combines the qualities of wisdom, upright dealing, and courage. The stoic will thus be immune from life's gyrations as he will be superior to the wounds from some of life's dirty tricks. But things can be carried to the extreme; the stern Cato found it beneath him to have human feelings. A more human version can be read in Seneca's *Letters from a Stoic*, a soothing and surprisingly readable book that I distribute to my trader friends (Seneca also took his own life when cornered by destiny).

RANDOMNESS AND PERSONAL ELEGANCE

The reader knows my opinion on unsolicited advice and sermons on how to behave in life. Recall that ideas do not truly sink in when emotions come into play; we do not use our rational brain outside of classrooms. Self-help books (even when they are not written by charlatans) are largely ineffectual. Good, enlightened (and "friendly") advice and eloquent sermons do not register for more than a few moments when they go against our wiring. The interesting thing about stoicism is that it plays on dignity and personal aesthetics, which are part of our genes. Start stressing personal elegance at your next misfortune. Exhibit *sapere vivere* ("know how to live") in all circumstances.

Dress at your best on your execution day (shave carefully); try to leave a good impression on the death squad by standing erect and proud. Try not to play victim when diagnosed with cancer (hide it from others and only share the information with the doctor—it will avert the platitudes and nobody will treat you like a victim worthy of their pity; in addition, the dignified attitude will make both defeat and victory feel equally heroic). Be extremely courteous to your assistant when you lose money (instead of taking it out on him as many of the traders whom I scorn routinely do). Try not to blame others for your fate, even if they deserve blame. Never exhibit any self-pity, even if your significant other bolts with the handsome ski instructor or the younger aspiring model. Do not complain. If you suffer from a benign version of the "attitude problem," like one of my childhood friends, do not start playing nice guy if your business dries up (he sent a heroic e-mail to his colleagues informing them "less business, but same attitude"). The only article Lady Fortuna has no control over is your behavior. Good luck.

•

SOLON TOLD YOU SO

Beware the London Traffic Jams

A couple of years after we left him looking at John smoking a cigarette with a modicum of *schadenfreude*, Nero's skepticism ended up paying off. Simultaneously as he beat the 28% odds, up to the point of complete cure, he made a series of exhilarating personal and professional victories. Not only did he end up sampling the next level of wealth but he got the riches right when other Wall Street hotshots got poor, which could have allowed him to buy the goods they owned at very large discounts, if he wanted to. But he acquired very little, and certainly none of the goods Wall Streeters usually buy. But Nero did engage in occasional excess.

Friday afternoon traffic in London can be dreadful. Nero started spending more time there. He developed an obsession with traffic jams. One day he spent five hours moving west from his office in the city of London toward a cottage in the Cotswolds, where he stayed most weekends. The frustration prompted Nero to get a helicopter-flying license, through a crash course in Cambridgeshire. He realized that the train was probably an easier solution to get out of town for the weekend, but he felt the urge for a pet extravagance. The other result of his frustration was his no less dangerous commuting on a bicycle between his flat in Kensington and his office in the city.

Nero's excessive probability-consciousness in his profession somehow did not register fully into his treatment of physical risk. For Nero's helicopter crashed as he was landing it near Battersea Park on a windy day. He was alone in it. In the end the black swan got its man.

•

THREE AFTERTHOUGHTS
IN THE SHOWER

Owing to the subject's tentacles and its author's ruminating nature, this book keeps growing like a living object. I will add in this section a few post-thoughts I've had in the shower and in the few boring philosophy lectures I've attended (without wanting to offend my new colleagues in the thinking business, I discovered that listening to a speaker reciting *verbatim* his lecture notes makes me invariably daydream).

FIRST THOUGHT:
THE INVERSE SKILLS PROBLEM

The higher up the corporate ladder, the higher the compensation to the individual. This might be justified, as it makes plenty of sense to pay individuals according to their contributions. However, and in general (provided we exclude risk-bearing entrepreneurs), the higher up the corporate ladder, the *lower* the evidence of such contribution. I call this the *inverse rule*.

I will be deriving the point by mere logical arguments. Chapter 2 made the distinction between those skills that are visible (like the abilities of a dentist) and those that present more difficulty in nailing down, especially when the subject belongs to a randomness-laden profession (say, one that includes the occasional practice of Russian roulette). The degree of randomness in such an activity and our ability to isolate the contribution of the individual determine the visibility of the skills content. Accordingly, the cook at the company headquarters or the factory worker will exhibit their direct abilities with minimal uncertainty. These contributions may be modest but they are clearly definable. A patently incompetent professional cook who cannot distinguish salt from sugar or who tends to systematically overcook the meat would be easily caught, provided the diners have functioning taste buds. And if he gets it right by luck once, it also will be hard for him to get it right by sheer chance a second, third, and a thousandth time.

Repetitiveness is key for the revelation of skills because of what I called *ergodicity* in Chapter 8—the detection of long-term properties, particularly when these exist. If you bang one million dollars at your next visit to Las Vegas at the roulette table in one single shot, you will not be able to ascertain from this single outcome whether the house has the advantage or if you were particularly out of the gods' favor. If you slice your gamble into a series of one million bets of one dollar each, the amount you recover will systematically show the casino's advantage. This is the core of sampling theory, traditionally called the *law of large numbers*.

To view it in another way, consider the difference between judging *on process* and judging *on results*. Lower-ranking persons in the enterprise are judged on both process and results—in fact, owing to the repetitive aspect of their efforts, their process converges rapidly to results. But top management is only paid on result—no matter the process. There seems to be no such thing as a foolish decision if it results in profits. "Money talks," we are often told. The rest is supposed to be philosophy.

Now take a peek inside the chief executive suite. Clearly, the decisions there are not repeatable. CEOs take a small number of large decisions, more like the person walking into the casino with a single million-dollar bet. External factors, such as the environment, play a considerably larger role than with the cook. The link between the skill of the CEO and the results of the company are tenuous. By some argument, the boss of the company may be unskilled labor but one who presents the necessary attributes of charisma and the package that makes for good MBA talk. In other words, he may be subjected to the monkey-on-the-typewriter problem. There are so many companies doing all kinds of things that some of them are bound to make "the right decision."

It is a very old problem. It is just that, with the acceleration of the power law–style winner-takes-all effects in our environment, such differences in outcomes are more accentuated, more visible, and more offensive to people's sense of fairness. In the old days, the CEO was getting ten to twenty times what the janitor earned. Today, he can get several thousand times that.

I am excluding entrepreneurs from this discussion for the obvious reason: These are people who stuck their necks out for some idea, and risked belonging to the vast cemetery of those who did not make it. But CEOs are not entrepreneurs. As a matter a fact, they are often *empty suits*. In the "quant" world, the designation *empty suit* applies to the category of persons who are good at looking the part but nothing more. More appropriately, what they have is skill in getting promoted within a company rather than pure skills in making optimal decisions—we call that "corporate political skill." These are people mostly trained at using PowerPoint presentations.

There is an asymmetry, as these executives have almost nothing to lose. Assume that two equally charismatic, empty-suit-style twin brothers manage to climb the corporate ladder to get two different jobs in two different corporations. Assume that they own good-looking suits, that they have MBAs, and that they are tall (the only truly visible predictor of corporate success is to be taller than average). They flip coins in secret and randomly take completely opposite actions, leading to great failure for one and great success for the other. We end up with a mildly wealthy, but fired, executive and his extremely wealthy, and still operating, twin brother. The shareholder bore the risk; the executives got the reward.

The problem is as old as leadership. Our attribution of heroism to those who took crazy decisions but were lucky enough to win shows the aberration—we continue to worship those who won battles and despise those who lost, no matter the reason. I wonder how many historians use luck in their interpretation of success—or how many are conscious of the difference between process and result.

I insist that it is not society's problem but that of the investors. If shareholders are foolish enough to pay someone $200 million to just wear a good-looking suit and ring a bell, as they did with the New York Stock Exchange's Richard Grasso in 2003, it is their own money they part with, not yours and mine. It is a corporate governance issue.

The situation is not much better in a bureaucratic economy. Outside the capitalistic system, presumed talent flows to the governmental positions, where the currency is prestige, power, and social rank. There, too, it is distributed disproportionately. The contributions of civil servants might be even more difficult to judge than those of the executives of a corporation—and the scrutiny is smaller. The central banker lowers interest rates, a recovery ensues, but we do not know whether he caused it or if he slowed it down. We can't even know that he didn't destabilize the economy by increasing the risk of future inflation. He can always fit a theoretical explanation, but economics is a narrative discipline, and explanations are easy to fit retrospectively.

The problem may not be incurable. It is just that we need to drill into the heads of those who measure the contribution of executives that what they see is not necessarily what is there. Shareholders, in the end, are the ones who are fooled by randomness.

SECOND THOUGHT: ON SOME ADDITIONAL BENEFITS OF RANDOMNESS

Uncertainty and Happiness

Have you ever had a weeknight dinner in New York City with a suburban commuter? Odds are that the shadow of the schedule will be imprinted in his consciousness. He will be tightly aware of the clock, pacing his meal in such a way that he does not miss the 7:08 because after that one, there are no more express trains and he would be reduced to taking the 7:42 local, something that appears to be very undesirable. He will cut the conversation short around 6:58, offer a quick handshake, then zoom out of the restaurant to catch his train with maximal efficiency. You will also be stuck with the bill. Since the meal is not finished, and the bill is not ready, your manners will force you to tell him that it's on you. You will also finish the cup of decaffeinated skim cappuccino all alone while staring at his empty seat and wondering why people get trapped by choice into such a life.

Now deprive him of his schedule—or randomize the time of departures of the trains so they no longer obey a fixed and known timetable. Given that what is random and what you do not know are functionally the same, you do not have to ask the New York area Metropolitan Transit Authority to randomize their trains for the purpose of the experiment: Just assume that he is deprived of knowledge of the various departure times. All he would know is that they operate about every, say, thirty-five minutes. What would he do under such a scenario? Although you might still end up paying for dinner, he would let the meal follow its natural course, then leisurely walk to the nearby station, where he would have to wait for the next train to show up. The time difference between the two situations will be a little more than a quarter of an hour. Another way to see the contrast between a known and an unknown schedule is to compare his condition to that of another diner who has to use the subway to go home, for an equivalent distance, but without a known and fixed schedule. Subway riders are freer of their schedule, and not just because of the higher frequency of trains. Uncertainty protects them from themselves.

Chapter 10 showed, with the illustration of Buridan's donkey, that randomness is not always unwelcome. This discussion aims to show how some degree of unpredictability (or lack of knowledge) can be beneficial to our defective species. A slightly random schedule prevents us from optimizing and being exceedingly efficient, particularly in the wrong things. This little bit of uncertainty might make the diner relax and forget the time pressures. He would be forced to act as a *satisficer* instead of a *maximizer* (Chapter 11 discussed Simon's satisficing as a blend of satisfying and maximizing)—research on happiness shows that those who live under a self-imposed pressure to be optimal in their enjoyment of things suffer a measure of distress.

The difference between satisficers and optimizers raises a few questions. We know that people of a happy disposition tend to be of the satisficing kind, with a set idea of what they want in life and an ability to stop upon gaining satisfaction. Their goals and desires do not move along with the experiences. They do not tend to experience the internal treadmill effects of constantly trying to improve on their consumption of goods by seeking higher and higher levels of sophistication. In other words, they are neither avaricious nor insatiable. An optimizer, by comparison, is the kind of person who will uproot himself and change his official residence just to reduce his tax bill by a few percentage points. (You would think that the entire point of a higher income is to be free to choose where to live; in fact it seems, for these people, wealth causes them to increase their dependence!) Getting rich results in his seeing flaws in the goods and services he buys. The coffee is not warm enough. The cook no longer deserves the three stars given to him by the Michelin guide (he will write to the editors). The table is too far from the window. People who get promoted to important positions usually suffer from tightness of schedules: Everything has an allotted time. When they travel, everything is "organized" with optimizing intent, including lunch at 12:45 with the president of the company (a table not too far from the window), the Stairmaster at 4:40, and opera at 8:00.

Causality is not clear: The question remains whether optimizers are unhappy because they are constantly seeking a better deal or if unhappy people tend to optimize out of their misery. In any case, randomness seems to operate either as a cure or as Novocain!

I am convinced that we are not made for clear-cut, well-delineated schedules. We are made to live like firemen, with downtime for lounging and meditating between calls, under the protection of protective uncertainty. Regrettably, some people might be involuntarily turned into optimizers, like a suburban child having his weekend minutes squeezed between karate, guitar lessons, and religious education. As I am writing these lines I am on a slow train in the Alps, comfortably shielded from traveling businesspersons. People around me are either students or retired persons, or those who do not have "important appointments," hence not afraid of what they call wasted time. To go from Munich to Milan, I picked the seven-and-a-half-hour train instead of the plane, which no self-respecting businessperson would do on a weekday, and am enjoying an air unpolluted by persons squeezed by life.

I came to this conclusion when, about a decade ago, I stopped using an alarm clock. I still woke up around the same time, but I followed my own personal clock. A dozen minutes of fuzziness and variability in my schedule made a considerable difference. True, there are some activities that require such dependability that an alarm clock is necessary, but I am free to choose a profession where I am not a slave to external pressure. Living like this, one can also go to bed early and not optimize one's schedule by squeezing every minute out of one's evening. At the limit, you can decide whether to be (relatively) poor, but free of your time, or rich but as dependent as a slave.

It took me a while to figure out that we are not designed for schedules. The realization came when I recognized the difference between writing a paper and writing a book. Books are fun to write, papers are painful. I tend to find the activity of writing greatly entertaining, given that I do it without any external constraint. You write, and may interrupt your activity, even in midsentence, the second it stops being attractive. After the success of this book, I was asked to write papers by the editors of a variety of professional and scientific journals. Then they asked me how long the piece should be. What? How long? For the first time in my life, I experienced a loss of pleasure in writing! Then I figured out a personal rule: For writing to be agreeable to me, *the length of the piece needs to remain unpredictable.* If I see the end of it, or if I am subjected to the shadow of an outline, I give up. I repeat that our ancestors were not subjected to outlines, schedules, and administrative deadlines.

Another way to see the beastly aspect of schedules and rigid projections is to think in limit situations. Would you like to know with great precision the date of your death? Would you like to know who committed the crime before the beginning of the movie? Actually, wouldn't it be better if the length of movies were kept a secret?

The Scrambling of Messages

Besides its effect on well-being, uncertainty presents tangible informational benefits, particularly with the scrambling of potentially damaging, and self-fulfilling, messages. Consider a currency pegged by a central bank to a fixed rate. The bank's official policy is to use its reserves to support it by buying and selling its currency in the open market, a procedure called *intervention*. But should the currency rate drop a tiny bit, people will immediately get the message that the intervention failed to support the currency and that the devaluation is coming. A pegged currency is not supposed to fluctuate; the slightest downward fluctuation is meant to be a harbinger of bad news! The rush to sell would cause a self-feeding frenzy leading to certain devaluation.

Now consider an environment where the central bank allows some noise around the official band. It does not promise a fixed rate, but one that can fluctuate a bit before the bank starts intervening. A small drop would not be considered to bear much information. The existence of noise leads us to avoid reading too much into variations. *Fluctuat nec mergitur* (it fluctuates but does not sink).

This point has applications in evolutionary biology, evolutionary game theory, and conflict situations. A mild degree of unpredictability in your behavior can help you to protect yourself in situations of conflict. Say you always have the same threshold of reactions. You take a set level of abuse, say seventeen insulting remarks per week, before getting into a rage and punching the eighteenth offender in the nose. Such predictability will allow people to take advantage of you up to that well-known trigger point and stop there. But if you randomize your trigger point, sometimes overreacting at the slightest joke, people will not know in advance how far they can push you. The same applies to governments in conflicts: They need to convince their adversaries that they are crazy enough to sometimes overreact to a small peccadillo. Even the magnitude of their reaction should be hard to foretell. Unpredictability is a strong deterrent.

THIRD THOUGHT: STANDING ON ONE LEG

I have been periodically challenged to compress all this business of randomness into a few sentences, so even an MBA can understand it (surprisingly, MBAs, in spite of the insults, represent a significant portion of my readership, simply because they think that my ideas apply to other MBAs and not to them).

This brings to mind Rabbi Hillel's story, when he was asked by someone particularly lazy if Hillel could teach him the Torah while the student was standing on one leg. Rabbi Hillel's genius is that he did not *summarize;* instead, he provided the core generator of the idea, the axiomatic framework, which I paraphrase as follows: *Don't do to others what you don't want them to do to you; the rest is just commentary.*

It took me an entire lifetime to find out what my generator is. It is: *We favor the visible, the embedded, the personal, the narrated, and the tangible; we scorn the abstract.* Everything good (aesthetics, ethics) and wrong (Fooled by Randomness) with us seems to flow from it.

ACKNOWLEDGMENTS
FOR THE FIRST EDITION

First, I would like to thank friends who can be considered rightful coauthors. I am grateful to New York intellectual and expert in randomness Stan Jonas (I do not know any other designation that would do him justice) for half a lifetime of conversations into all subjects bordering on probability with the animation and the zeal of the neophyte. I thank my probabilist friend Don Geman (husband of Helyette Geman, my thesis director) for his enthusiastic support for my book; he also made me realize that probabilists are born, not made—many mathematicians are capable of computing, but not understanding, probability (they are no better than the general population in exerting probabilistic judgments). The real book started with an all-night conversation with my erudite friend Jamil Baz during the summer of 1987, as he discussed the formation of "new" and "old" money among families. I was then a budding trader and he scorned the arrogant Salomon Brothers traders who surrounded him (he was proved right). He instilled in me the voracious introspection about my performance in life and really gave me the idea for this book. Both of us ended up getting doctorates later in life, on an almost identical subject matter. I have also dragged many people on (very long) walks in New York, London, or Paris, discussing some parts of this book, such as the late Jimmy Powers, who helped nurture my trading early on, and who kept repeating "anyone can buy and sell," or my encyclopedic friend David Pastel, equally at ease with literature, mathematics, and Semitic languages. I have also engaged my lucid Popperian colleague Jonathan Waxman in numerous conversations on the integration of Karl Popper's ideas into our life as traders.

Second, I have been lucky to meet Myles Thompson and David Wilson, when they both were at J. Wiley & Sons. Myles understood that books need not be written to satisfy a predefined labeled audience, but that a book will find its own unique set of readers—thus giving more credit to the reader than the off-the-rack publisher. As to David, he believed enough in the book to push me to take it into its natural course, free of all labels and taxonomies. David saw me the way I view myself: someone who has a passion for probability and randomness, who is obsessed with literature but happens to be a trader, rather than a generic "expert." He also saved my idiosyncratic style from the dulling of the editing process (for all its faults, the style is mine). Finally, Mina Samuels proved to be the greatest conceivable editor: immensely intuitive, cultured, aesthetically concerned, yet nonintrusive.

Many friends have fed me with ideas during conversations, ideas that found their way into the text. I can mention the usual suspects, all of them prime conversationalists: Cynthia Shelton Taleb, Helyette Geman, Marie-Christine Riachi, Paul Wilmott, Shaiy Pilpel, David DeRosa, Eric Briys, Sid Kahn, Jim Gatheral, Bernard Oppetit, Cyrus Pirasteh, Martin Mayer, Bruno Dupire, Raphael Douady, Marco Avellaneda, Didier Javice, Neil Chriss, and Philippe Asseily.

Some of these chapters were composed and discussed as part of the "Odeon Circle," as my friends and I met with a varying degree of regularity (on Wednesdays at 10 p.m. after my Courant class) at the bar of the restaurant Odeon in Tribeca. Genius loci ("the spirit of the place") and outstanding Odeon staff member Tarek Khelifi made sure that we were well taken care of and enforced our assiduity by making me feel guilty on no-shows, thus helping greatly with the elaboration of the book. We owe him a lot.

I must also acknowledge the people who read the MS, diligently helped with the errors, or contributed to the elaboration of the book with useful comments: Inge Ivchenko, Danny Tosto, Manos Vourkoutiotis, Stan Metelits, Jack Rabinowitz, Silverio Foresi, Achilles Venetoulias, and Nicholas Stephanou. Erik Stettler was invaluable in his role as a shadow copy editor. All mistakes are mine.

Finally, many versions of this book sat on the Web, yielding sporadic (and random) bursts of letters of encouragement, corrections, and valuable questions, which made me weave answers into the text. Many chapters of this book came in response to readers' questions. Francesco Corielli from Bocconi alerted me on the biases in the dissemination of scientific results.

This book was written and finished after I founded Empirica, my intellectual home, "Camp Empirica," in the woods in the back country of Greenwich, Connecticut, which I designed to fit my taste and feel like a hobby: a combination of an applied probability research laboratory, athletic summer camp, and, not least, a trading operation (I had experienced one of my best professional years while writing these lines). I thank all the like-minded people who helped fuel the stimulating atmosphere there: Pallop Angsupun, Danny Tosto, Peter Halle, Mark Spitznagel, Yuzhao Zhang, and Cyril de Lambilly as well as the members of Paloma Partners such as Tom Witz, who challenged our wisdom on a daily basis, and Donald Sussman, who supplied me with his penetrating judgment.

A TRIP TO THE LIBRARY

Notes and Reading Recommendations

NOTES

I confess that, as a practitioner of randomness, I focused primarily on the defects of my *own* thinking (and that of a few people I've observed or tracked through time). I also intended the book to be playful, which is not very compatible with referencing every idea to some scientific paper to give it a degree of respectability. I take the liberty in this section to finesse a few points and to provide select references (of the "further reading" variety)—but references linked to matters that I directly experienced. I repeat that this is a personal essay, not a treatise.

On completion of this compilation I discovered the predominance of matters relating to human nature (mostly empirical psychology) over things mathematical. Sign of the times: I am convinced that the next edition, hopefully two years from now, will have plenty of references and notes in neurobiology and neuroeconomics.

PREFACE

Hindsight bias: a.k.a Monday morning quarterback. See Fischhoff (1982).

Clinical knowledge: The problem of clinicians not knowing what they do not know, and not quite figuring it out. See Meehl (1954) for the seminal introduction. "It is clear that the dogmatic, complacent assertion sometimes heard from clinicians that 'naturally' clinical prediction, being based on 'real understanding' is superior, is simply not justified by the facts to date." In his testing, in all but one case, predictions made by actuarial means were equal to or better than clinical methods. Even worse: In a later paper, he changed his mind about that one exception. Since Meehl's work there has been a long tradition of examination of expert opinions, confirming the same results. This problem applies to about every profession—particularly journalists and economists. We will discuss in further notes the associated problem of self-knowledge.

Montaigne vs Descartes: I thank the artificial intelligence researcher and omnivorous reader Peter McBurney for bringing to my attention the discussion in Toulmin (1990). On that I have to make the sad remark that Descartes was originally a skeptic (as attested by his demon thought experiment) but the so-called Cartesian mind corresponds to someone with an appetite for certainties. Descartes' idea in its original form is that there are very few certainties outside of narrowly defined deductive statements, not that everything we think about needs to be deductive.

Affirming the consequent: The logical fallacy is generally presented as follows.

If p then q

q

Therefore, p

(All people in the Smith family are tall; he is tall therefore he belongs to the Smith family).

The track record of the general population in correctly making such inference is exceedingly poor. Although it is not customary to quote textbooks, I refer the reader to the excellent Eysenck and Keane (2000) for a list of the research papers on the different difficulties—up to 70% of the population can make such a mistake!

The millionaire mind: Stanley (2000). He also figured out (correctly) that the rich were "risk takers" and inferred (incorrectly) that risk taking made one rich. Had he examined the population of failed entrepreneurs he would have also inferred (correctly) that the failed entrepreneurs too were "risk takers."

Journalists are "practical": I heard at least four times the word *practical* on the part of journalists trying to justify their simplification. The television show that wanted me to present three stock recommendations wanted something "practical," not theories.

PROLOGUE

Mathematics conflicts with probability: One is about certainties, the other about the exact opposite. This explains the disrespect held by pure mathematicians for the subject of probability for a long time—and the difficulty in integrating the two. It is not until recently that it was termed "the logic of science"—the title of the posthumous Jaynes (2003). Interestingly, this book is also perhaps the most complete account of the mathematics of the subject—he manages to use probability as an expansion of conventional logic.

The prominent mathematician David Mumford, a Fields medalist, repents for his former scorn for probability. He writes in *The Dawning of the Age of Stochasticity* (Mumford, 1999): "For over two millennia, Aristotle's logic has ruled over the thinking of Western intellectuals. All precise theories, all scientific models, even models of the process of thinking itself, have in principle conformed to the straight-jacket of logic. But from its shady beginnings devising gambling strategies and counting corpses in medieval London, probability theory and statistical inference now emerge as better foundations for scientific models, especially those of the process of thinking and as essential ingredients of theoretical mathematics, even the foundations of mathematics itself. We propose that this sea change in our perspective will affect virtually all of mathematics in the next century."

Courage or foolishness: For an examination of that notion of "courage" and "guts," see Kahneman and Lovallo (1993). See also a discussion in Hilton (2003). I drew the idea from Daniel Kahneman's presentation in Rome in April 2003 (Kahneman, 2003).

Cognitive errors in forecasting: Tversky and Kahneman (1971), Tversky and Kahneman (1982), and Lichtenstein, Fischhoff and Phillips (1977).

Utopian/tragic: The essayist and prominent (scientific) intellectual Steven Pinker popularized the distinction (originally attributable to the political scholar Thomas Sowell). See Sowell (1987), Pinker (2002). Actually, the distinction is not so clear. Some people actually believe, for instance, that Milton Friedman is a utopist in the sense that all ills come from governments and that getting rid of government would be a great panacea.

Fallibility and infallibilism: Peirce (in a prospectus for a never written book), writes, "Nothing can be more completely contrary to a philosophy, the fruit of a scientific life, than infallibilism, whether arrayed in the old ecclesiastical trappings, or under its recent 'scientific' disguise." (Brent, 1993). For a brief and very readable acquaintance to the works of Peirce, Menand (2001). It draws on his sole biography, Brent (1993).

CHAPTER 1

Relative compared to absolute position: See Kahneman, Knetsch and Thaler (1986). Robert Frank is an interesting researcher who spent part of his career thinking about the problem of status, rank, and relative income: See Frank (1985), and the very readable Frank (1999). The latter includes discussions on the interesting proposer/responder problem where people forego windfall profits in order to deprive others of a larger share. One person proposes to the other a share of, say, $100. She can accept or refuse. If she refuses, both get nothing.

Even more vicious results have been shown by researchers who studied how much people would *pay* to lower other people's income: See Zizzo and Oswald (2001). On that also, see Burnham (2003) (he ran an experiment measuring the testosterone levels in economic exchange).

Serotonin and pecking order: Frank (1999) includes a discussion.

On the social role of the psychopath: See Horrobin (2002). While it may have some extreme views on the point, the book reviews discussions of the theories around the success realized by the psychopaths. Also, see Carter (1999) for a presentation of the advantage some people have in being separated from the feeling of empathy and compassion.

Social emotions: Damasio (2003): "One of the many reasons why people become leaders and others followers, why so many command respect, has little to do with knowledge or skills and a lot to do with how some physical traits and the manner of a given individual promote certain emotional responses in others."

Literature on emotions: For a review of the current scientific ideas, see the excellent compact Evans (2002). Evans belongs to the new breed of the philosopher/essayist contemplating large themes with a scientific mind. Elster (1998) goes into the broad social implications of emotions. The bestselling Goleman (1995) offers a surprisingly complete account (the fact that it is a bestseller is surprising: We are aware of our irrationality but it does not seem to help).

CHAPTER 2

Possible worlds: Kripke (1980).

Many worlds: See the excellently written Deutsch (1997). I also suggest a visit to the author's rich website. The earlier primary work can be found in DeWitt and Graham (1973), which contains Hugh Everett's original paper.

Economics of uncertainty and possible states of nature: See Debreu (1959). For a presentation of lattice state-space methods in mathematical finance, see Ingersoll (1987) (well structured though dry and very, very boring, like the personality of its author), and the more jargon-laden Huang and Litzenberger (1988). For an economics-oriented presentation, see Hirshleifer and Riley (1992).

For the works of Shiller: See Shiller (2000). The more technical work is in the (originally) controversial Shiller (1981). See also Shiller (1990). For a compilation: Shiller (1989). See also Kurz (1997) for a discussion of endogenous uncertainty.

Risk and emotions: Given the growing recent interest in the emotional role in behavior, there has been a growing literature on the role of emotions in both risk bearing and risk avoidance: The "risk as feeling" theory: See Loewenstein, Weber, Hsee and Welch (2001), and Slovic, Finucane,

Peters and MacGregor (2003a). For a survey, see Slovic, Finucane, Peters and MacGregor (2003b). See also Slovic (1987).

For a discussion of the affect heuristic: See Finucane, Alhakami, Slovic and Johnson (2000).

Emotions and cognition: For the effect of emotions on cognition, see LeDoux (2002).

Availability heuristic (how easily things come to mind): Tversky and Kahneman (1973).

Real incidence of catastrophes: For an insightful discussion, see Albouy (2002).

On sayings and proverbs: Psychologists have long examined the gullibility of people in social settings facing well-sounding proverbs. For instance, experiments since the 1960s have been made where people are asked whether they believed that a proverb is right, while another cohort is presented the opposite meaning. For a presentation of the hilarious results, see Myers (2002).

Epiphenomena: See the beautiful Wegner (2002).

CHAPTER 3

Keynes: Keynes' *Treatise on Probability* (Keynes, 1989, 1920) remains in many people's opinion the most important single work on the subject— particularly considering Keynes' youth at the time of composition (it was published years after he finished it). In it he develops the critical notion of subjective probability.

Les gommes: Robbe-Grillet (1985).

Pseudoscientific historicism: For an example, I suggest Fukuyama (1992).

Fears built into our genes: This is not strictly true—genetic traits need to be culturally activated. We are wired for some fears, such as fears of snakes, but monkeys who have never seen a snake do not have it. They need the sight of the fear in the facial features of another monkey to start getting scared (LeDoux, 1998).

Amnesia and risk avoidance: Damasio (2000) presents the case of David the amnesic patient who knew to avoid those who abused him. See also Lewis, Amini and Lannon (2000). Their book presents a pedagogic discussion of "camouflaged learning," in the form of implicit memory, as opposed to explicit memory (neocortical). The book portrays memory as a correlation in neuron connectivity rather than some CD-style recording— which explains the revisions of memory by people after events.

Why don't we learn from our past history?: Two strains of literature. (1) The recent "stranger to ourselves" line of research in psychology (Wilson 2002). (2) The literature on "immune neglect," Wilson, Meyers and Gilbert (2001) and Wilson, Gilbert and Centerbar (2003). Literally, people don't learn from their past reactions to good and bad things.

Literature on bubbles: There is a long tradition, see Kindleberger (2001), MacKay (2002), Galbraith (1991), Chancellor (1999), and of course Shiller (2000). Shiller with a little work may be convinced to do a second edition.

Long-term capital management: See Lowenstein (2000).

Stress and randomness: Sapolsky (1998) is a popular, sometimes hilarious presentation. The author specializes among other things on the effect of glucocorticoids released at times of stress on the atrophy of the hycocampus, hampering the formation of new memory and brain plasticity. More technical, Sapolsky (2003).

Brain asymmetries with gains/losses: See Gehring and Willoughby (2002). See the works of Davidson on the anterior brain asymmetry (a clear summary and popular presentation in Goleman 2003). See also Shizgal (1999).

The dentist and prospect theory: Kahneman and Tversky (1979). In this seminal discussion they present agents as interested in differences and resetting their pain/pleasure level at zero as "anchor." The gist of it is that "wealth" does not matter, almost only differences in wealth, since the resetting cancels the effect of the accumulation. Think of John hitting wealth of $1 million from below or above and the impact on his well-being. The difference between utility of wealth and utility of changes in wealth is not trivial: It leads to dependence on the observation period. In fact the notion, taken to its limit, leads to the complete revision of economic theory: Neoclassical economics will no longer be useful beyond mathematical exercises. There have been vigorous such discussions in the hedonistic literature as well: See Kahneman, Diener and Schwarz (1999).

CHAPTER 4

Public and scientific intellectual: Brockman (1995) offers presentations by the "who's who" in the new scientific intellectual tradition. See also his website, www.edge.org. For a physicist's position on the culture wars, Weinberg (2001). For a presentation of a public intellectual, see Posner (2002). Note that Florida Atlantic University offers a Ph.D. to become a public intellectual—literary, since scientists need no such artifice.

The hoax: Sokal (1996).

The Selfish Gene: Dawkins (1989, 1976). Hegel: In Popper (1994).

Exquisite cadavers: Nadeau (1970).

The generator: www.monash.edu.au.

Language and probability: There is a very large connection between language and probability; it has been studied by thinkers and scientists via the sister methods of entropy and information theory—one can reduce the dimensionality of a message by eliminating redundancy, for instance; what is left is measured as information content (think of zipping a file) and is linked to the notion of "entropy," which is the degree of disorder, the unpredictable that is left. Entropy is a very invasive notion as it relates to aesthetics and thermodynamics. See Campbell (1982) for a literary presentation, and Cover and Thomas (1991) for a scientific one, particularly the discussion on the "entropy of English." For a classic discussion of entropy and art, Arnheim (1971), though the connection between entropy and probability was not yet clear at the time. See Georgescu-Roegen (1971) for a (perhaps) pioneering discussion of entropy in economics.

CHAPTER 5

The firehouse effect and the convergence of opinions: There are plenty of discussions in the psychology literature of such convergence of opinions, particularly in the area of mate selection or what Keynes calls "the beauty contest," as people tend to choose what other people choose, causing positive-feedback loops.

An interesting manifestation is the autokinetic effect. When people gaze at a stationary light in a room they see it moving after a while and can estimate the amount of movement, not knowing that it is an optical illusion. When isolated the subjects give wildly varying speeds of movement; when tested in a group they converge to a common speed of movement: See Plotkin (1998). Sornette (2003) gives an interesting account of the feedback loops that result from herding written in light, but with extremely intuitive mathematics.

Biology of imitation: See Dugatkin (2001).

Evolution and small probabilities: Evolution is principally a probabilistic concept. Can it be fooled by randomness? Can the least skilled survive? There is a prevalent strain of Darwinism, called naive Darwinism, that believes that any species or member of a species that dominates at any point has been selected by evolution because they have an advantage over others. This results from a common misunderstanding of local and global optima, mixed with an inability to get rid of the belief in the law of small numbers (overinference from small data sets). Just put two people in a random environment, say a gambling casino, for a weekend. One of them will fare better than the other. To a naive observer the one who fares better will have a survival advantage over the other. If he is taller or has some trait that distinguishes him from the other, such trait will be identified by the naive observer as the explanation of the difference in fitness. Some people do it with traders—make them compete in a formal competition. Consider also the naive evolutionary thinking positing the "optimality" of selection—the founder of sociobiology does not agree with such optimality when it comes to rare events: E. O. Wilson (2002) writes: "The human brain evidently evolved to commit itself emotionally only to a small piece of geography, a limited band of kinsmen, and two or three generations into the future. To look neither far ahead nor far afield is elemental in a Darwinian sense. *We are innately inclined to ignore any distant possibility not yet requiring examination. It is, people say, just good common sense.* Why do they think in this shortsighted way? "The reason is simple: It is a hardwired part of our Paleolithic heritage. For hundreds of millennia, those who worked for short-term gain within a small circle of relatives and friends lived longer and left more offspring—even when their collective striving caused their chiefdoms and empires to crumble around them. The long view that might have saved their distant descendants required a vision and extended altruism instinctively difficult to marshal."

See also Miller (2000): "Evolution has no foresight. It lacks the long-term vision of drug company management. A species can't raise venture capital to pay its bills while its research team . . . Each species has to stay biologically profitable every generation, or else it goes extinct. Species always have cashflow problems that prohibit speculative investments in their future. More to the point, every gene underlying every potential innovation has to yield higher evolutionary payoffs than competing genes, or it will disappear before the innovation evolves any further. This makes it hard to explain innovations."

CHAPTER 6

Fooled by negative skewness: The first hint of an explanation for the popularity of negatively skewed payoffs comes from the early literature on behavior under uncertainty, with the "small number problem." Tversky and Kahneman (1971) write: "We submit that people view a sample randomly drawn from a population as highly representative, that is, similar to a population in all essential characteristics." The consequence is the inductive fallacy: overconfidence in the ability to infer general properties from observed facts, "undue confidence in early trends," the stability of observed patterns and deriving conclusions with more confidence attached to them than can be warranted by the data. Worst, the agent finds causal explanations or perhaps distributional attributes that confirm his undue generalization. It is easy to see that the "small numbers" get exacerbated with skewness since most of the time the observed mean will be different from the true mean and most of the time the observed variance will be lower than the true one. Now consider that it is a fact that in life, unlike a laboratory or a casino, we do not observe the probability distribution from which random variables are drawn: We only see the realizations of these random processes. It would be nice if we could, but it remains that we do not measure probabilities as we would measure the temperature or the height of a person. This means that when we compute probabilities from past data we are making assumptions about the skewness of the generator of the random series—all data is conditional upon a generator. In short, with skewed packages, the camouflage of the properties comes into play *and* we tend to believe what we see. Taleb (2004).

Philosopher sometimes playing scientist: Nozik (1993).

Hollywood economics: De Vany (2003).

People are sensitive to sign rather than magnitude: Hsee and Rottenstreich (2004).

Lucas critique: Lucas (1978).

CHAPTER 7

Niederhoffer's book: Niederhoffer (1997).

Goodman's riddle of induction: One can take the issue of induction into a more difficult territory with the following riddle. Say the market went up every day for a month. For many people of inductive taste it could confirm the theory that it is going up every day. But consider: It may confirm the theory that it goes up every day then crashes—what we are witnessing is not an ascending market but one that *ascends then crashes*. When one observes a blue object it is possible to say that one is observing something blue until time *t*, beyond which it is green—that such object is not blue but "grue." Accordingly, by such logic, the fact that the market went up all this time may confirm that it will crash tomorrow! It confirms that we are observing a rising-crashing market. See Goodman (1954).

Writings by Soros: Soros (1988).

Hayek: See Hayek (1945) and the prophetic Hayek (1994), first published in 1945.

Popper's personality: Magee (1997), and Hacohen (2001). Also an entertaining account in Edmonds and Eidinow (2001).

CHAPTER 8

The millionaire next door: Stanley (1996).

Equity premium puzzle: There is an active academic discussion of the "equity premium" puzzle, taking the "premium" here to be the outperformance of stocks in relation to bonds and looking for possible explanations. Very little consideration was given to the possibility that the premium may have been an optical illusion owing to the survivorship bias—or that the process may include the occurrence of black swans. The discussion seems to have calmed a bit after the declines in the equity markets after the events of 2000–2002.

CHAPTER 9

Hot-hand effect: Gilovich, Vallone and Tversky (1985).

Stock analysts fooled by themselves: For a comparison between analysts and weather forecasters, see Taszka and Zielonka (2002).

Differences between returns: See Ambarish and Siegel (1996). The dull presenter was actually comparing "Sharpe ratios," i.e., returns scaled by their standard deviations (both annualized), named after the financial economist William Sharpe, but the concept has been commonly used in statistics and called "coefficient of variation." (Sharpe introduced the concept in the context of the normative theory of asset pricing to compute the expected portfolio returns given some risk profile, not as a statistical device.) Not counting the survivorship bias, over a given twelve-month period, assuming (very generously) the Gaussian distribution, the "Sharpe ratio" differences for two uncorrelated managers would exceed 1.8 with close to 50% probability. The speaker was discussing "Sharpe ratio" differences of around .15! Even assuming a five-year observation window, something very rare with hedge fund managers, things do not get much better.

Value of the seat: Even then, by some attribution bias, traders tend to believe that their income is due to their skills, not the "seat," or the "franchise" (i.e., the value of the order flow). The seat has a value as the New York Stock Exchange specialist "book" is worth quite large sums: See Hilton (2003). See also Taleb (1997) for a discussion of the time and place advantage.

Data mining: Sullivan, Timmermann and White (1999).

Dogs not barking: I thank my correspondent Francesco Corielli from Bocconi for his remark on meta-analysis.

CHAPTER 10

Networks: Arthur (1994). See Barabasi (2002), Watts (2003).

Nonlinear dynamics: For an introduction to nonlinear dynamics in finance, see Brock and De Lima (1995), and Brock, Hsieh and LeBaron (1991). See also the recent, and certainly the most complete, Sornette (2003). Sornette goes beyond just characterizing the process as fattailed and saying that the probability distribution is different from the one we learned in Finance 101. He studies the transition points: Say a book's sales become close to a critical point from which they will really take off. Their dynamics, conditional on past growth, become predictable.

The Tipping Point: Gladwell (2000). In the article that preceded the book (Gladwell, 1996) he writes: "The reason this seems surprising is that human beings prefer to think in linear terms. . . . I can remember struggling with these same theoretical questions as a child, when I tried to pour ketchup on my dinner. Like all children encountering this problem for the first time, I assumed that the solution was linear: That steadily increasing hits on the base of the bottle would yield steadily increasing amounts of ketchup out the other end. Not so, my father said, and he recited a ditty that, for me, remains the most concise statement of the fundamental nonlinearity of everyday life: 'Tomato ketchup in a bottle—None will come and then the lot'll.' "

Pareto: Before we had a generalized use of the bell curve, we took the ideas of Pareto with his distribution more seriously—its mark is the contribution of large deviations to the overall properties. Later elaborations led to the so-called Pareto-Levy or Levy-Stable distributions with (outside of special cases) some quite vicious properties (no known error rate). The reasons economists never liked to use it is that it does not offer tractable properties—economists like to write papers in which they offer the illusion of solutions, particularly in the form of mathematical answers. A Pareto-Levy distribution does not provide them with such luxury. For economic discussions on the ideas of Pareto, see Zajdenweber (2000), Bouvier (1999). For a presentation of the mathematics of Pareto-Levy distributions, see Voit (2001), and Mandelbrot (1997). There is a recent rediscovery of power law dynamics. Intuitively a power law distribution has the following property: If the power exponent were 2, then there would be 4 times more people with an income higher than $1 million than people with $2 million. The effect is that there is a very small probability of having an event of an extremely large deviation. More generally given a deviation x, the incidence of a deviation of a multiple of x will be that multiple to a given power exponent. The higher the exponent the lower the probability of a large deviation.

Spitznagel's remark: In Gladwell (2002).

Don't take "correlation" and those who use the word seriously: The same "A." of the lighter-throwing variety taught me a bit about the fallacy of the notion of correlation. "You do not seem to be correlated to anything" is the most common blame I've received when carrying my strategy of shooting for rare events. The following example might illustrate it. A nonlinear trading instrument, such as a put, will be positively correlated to the underlying security over many sample paths (say the put expires worthless in a bear market as the market did not drop enough), except of course upon becoming in the money and crossing the strike, in which case the correlation reverses with a vengeance. The reader should do himself a favor by not taking the notion of correlation seriously except in very narrow matters where linearity is justified.

CHAPTER 11

Probability "blindness": I borrow the expression from Piattelli-Palmarini (1994).

Discussion of "rationality": The concept is not so easy to handle. As the concept has been investigated in plenty of fields, it has been developed the most by economists as a normative theory of choice. Why did the economists develop such an interest in it? The basis of economic analysis is a concept of human nature and rationality embodied in the notion of *homo economicus*. The characteristics and behavior of such *homo economicus* are built into the postulates of consumer choice and include nonsatiation (more is *always* preferred to less) and transitivity (global consistency in choice). For instance, Arrow (1987) writes, "It is noteworthy that the everyday usage of the term 'rationality' does not correspond to the economist's definition as transitivity and completeness, that is maximization of something. The common understanding is instead the complete exploitation of information, sound reasoning, and so forth."

Perhaps the best way to see it for an economist is the maximization leading to a unique solution.

Even then, it is not easy. Who is maximizing what? To begin, there is a conflict between collective and individual rationality ("tragedy of the commons" seen by Keynes in his parable of the stadium where one's optimal strategy is to stand up, but collectively the optimal strategy is for everyone to remain seated). Another problem is seen in Arrow's voter's impossibility theorem. Consider also the following voter problem: People vote but the probability adjusted gains from voting can be less than the effort expended in going to the polling place. See Luce and Raiffa (1957) for a discussion of these paradoxes.

Note that the literature on rational choice under uncertainty is very extensive, cutting across fields, from evolutionary game theory to political science. But as John Harsanyi put it bluntly, *It is normative, and meant to be so.* This is a heroic statement: Saying that economics has abandoned its scientific pretensions and accepted that it does not describe how people *do* act but rather how they *should* act. It means that it has entered the realm of something else: philosophy (though not quite ethics). As such, an individual can accept it fully and should aim to act like the neoclassical man. If he can.

Ultimate/proximate as a solution to some rationality problems: Evolutionary theorists distinguish between proximate and ultimate cause.

Proximate cause: I eat *because* I am hungry.
Ultimate cause: If I didn't have an incentive to eat I would have gracefully exited the gene pool.

Now, if one invokes ultimate causes, plenty of behavior deemed locally irrational (like the voter problem above) can be interpreted as rational. It explains altruism: Why would you take a small risk to help a stranger from drowning? Visibly this impetus to help put us where we are today.

See Dawkins (1989, 1976) and Pinker (2002) for additional insights on the difference.

Rationality and scientism: Under the suggestion of my correspondent Peter McBurney I discovered the novel *We* by Yevgeny Zamyatin, a satire on Leninist Russia written in the 1920s and set in the far distant future, at a time when Taylorist and rationalist ideas had succeeded, apparently, in eliminating all uncertainty and irrationality from life.

Bounded rationality: Simon (1956), Simon (1957), Simon (1987a), and Simon (1987b).

Birth of the neurobiology of rationality: Berridge (2003) introduces a neurobiological dimension to rationality using two of Daniel Kahneman's four utilities (the experienced, remembered, predicted, and decision utilities) and setting irrationality if the decision utility exceeds the predicted one. There is a neural dimension to such irrationality: dopamine activity in the mesolimbic brain.

Compilation of the heuristics and biases papers in four volumes: Kahneman, Slovic and Tversky (1982), Kahneman and Tversky (2000), Gilovich, Griffin and Kahneman (2002), and Kahneman, Diener and Schwarz (1999).

Two systems of reasoning: See Sloman (1996), and Sloman (2002). See the summary in Kahneman and Frederick (2002). For the affect heuristic, see Zajonc (1980), and Zajonc (1984).

Evolutionary psychology/sociobiology: The most readable is Burnham and Phelan (2000). See Kreps and Davies (1993) for the general framework of ecology as optimization. See also Wilson (E. O., 2000), Winston (2002), the cartoons of Evans and Zarate (1999), Pinker (1997), and Burnham (1997).

Modularity: For the seminal work, see Fodor (1983) in philosophy and cognitive science, Cosmides and Tooby (1992) in evolutionary psychology.

The Wason selection task (written about in nearly every book on evolutionary psychology) is as follows. Consider the following two tests:

Problem 1: Suppose that I have a pack of cards, each of which has a letter written on one side and a number written on the other side. Suppose in addition that I claim that the following rule is true: *If a card has a vowel on one side, then it has an even number on the other side.* Imagine that I now show you four cards from the pack: E 6 K 9. Which card or cards should you turn over in order to decide whether the rule is true or false?

Problem 2: You are a bartender in a town where the legal age for drinking is twenty-one and feel responsible for the violations of the rules. You are confronted with the following situations and would have to ask the patron to show you either his age or what he is drinking. Which of the four patrons would you have to question?

1, drinking beer; 2, over twenty-one; 3, drinking Coke; 4, under 21.

While the two problems are identical (it is clear that you need to check only the first and last of the four cases) the majority of the population gets the first one wrong and the second one right. Evolutionary psychologists believe that the defects in solving the first problem and ease in the second show evidence of a cheater detection module—just consider that we adapted to the enforcement of cooperative tasks and are quick at identifying free riders.

Criteria of modularity: I borrow from the linguist Elisabeth Bates' presentation (Bates, 1994) of Fodor's nine criteria of modularity (ironically Bates is a skeptic on the subject). The information-processing criteria are: encapsulation (we cannot interfere with the functioning of a module), unconsciousness, speed (that's the point of the module), shallow outputs (we have no idea of the intermediate steps), and obligatory firing (a module generates predetermined outputs for predetermined inputs). The biological criteria that distinguish them from learned habits are: ontogenetic universals (they develop in characteristic sequence), localization (they use dedicated neural systems), and pathological universals (modules have characteristic pathologies across populations). Finally, modularity's most important property is its domain specificity.

Books on the physical brain: For the hierarchy reptilian/limbic/neocortical, see causal descriptions in Ratey (2001), Ramachandran and Blakeslee (1998), Carter (1999), Carter (2002), Conlan (1999), Lewis, Amini, and Lannon (2000), and Goleman (1995).

Emotional Brain: Damasio (1994) and LeDoux (1998). Bechara, Damasio, Damasion, and Tranel (1994) show the degradation of the risk-avoidance behavior of patients with damage in their ventromedial frontal cortex, a part of the brain that links us to our emotions. Emotions seem to play a critical role both ways. For the new field of neuroeconomics, see discussions in Glimcher (2002) and Camerer, Loewenstein and Prelec (2003).

Sensitivity to losses: Note that losses matter more than gains, but you become rapidly desensitized to them (a loss of $10,000 is better than ten losses of $1,000). Gains matter less than losses, and large gains even less (ten gains of $1,000 are better than one gain of $10,000).

Hedonic treadmill: My late friend Jimmy Powers used to go out of his way to show me very wealthy investment bankers acting miserably after a bad day. How good is all this wealth for them if they adjust to it to such a point that a single bad day can annihilate the effect of all these past successes? If things do not accumulate well then it follows that humans should follow a different set of strategies. This "resetting" shows the link to prospect theory.

Debate: Gigerenzer (1996), Kahneman and Tversky (1996), and Stanovich and West (2000). The evolutionary theorists are deemed to hold a Panglossian view: Evolution solves everything. Strangely, the debate is bitter not because of large divergences of opinions but because of small ones. *Simple Heuristics That Make Us Smart* is the title of a compilation of articles by Gigerenzer and his peers (Gigerenzer, 2000). See also Gigerenzer, Czerlinski and Martignon (2002).

Medical example: Bennett (1998). It is also discussed in Gigerenzer, Czerlinski and Martignon (2002). The heuristics and biases catalogue it as the base rate fallacy. The evolutionary theorists split into domain general (unconditional probability) as opposed to domain specific (conditional).

Behavioral finance: See Schleifer (2000) and Shefrin (2000) for a review. See also Thaler (1994b) and the original Thaler (1994a).

Domain-specific adaptations: Our lungs are a domain-specific adaptation meant to extract oxygen from the air and deposit it into our blood; they are not meant to circulate blood. For evolutionary psychologists the same applies to psychological adaptations.

Opaque process: For psychologists in the heuristics and biases tradition, System 1 is opaque, that is, not self-aware. This resembles the encapsulation and unconsciousness of modules discussed earlier.

Flow: See Csikszentmihalyi (1993) and Csikszentmihalyi (1998). I am quoting both to be safe but I don't know if there are differences between the books: The author seems to rewrite the same global idea in different ways.

Underestimation of possible outcomes: Hilton (2003).

The neurobiology of eye contact: Ramachandran and Blakeslee (1998) on the visual centers that project to the amygdala: "Scientists recording cell responses in the amygdala found that, in addition to responding to facial expressions and emotions, the cells also respond to the direction of eye gaze. For instance, one cell may fire if another person is looking directly at you, whereas a neighboring cell will fire only if that person's gaze is averted by a fraction of an inch. Still other cells fire when the gaze is way off to the left or the right. This phenomenon is not surprising given the important role that gaze direction plays in primate social communications—the averted gaze of guilt, shame or embarrassment; the intense, direct gaze of a lover, or the threatening stare of an enemy."

CHAPTER 12

Pigeons in a box: Skinner (1948).

Illusion of knowledge: Barber and Odean (2001) presents a discussion of the literature on the tendency to make a stronger inference than warranted by the data, which they call "Illusion of Knowledge."

CHAPTER 13

Arabic skeptics: Al-Ghazali (1989).

Rozan's book: Rozan (1999).

Mental accounting: Thaler (1980) and Kahneman, Knetch and Thaler (1991).

Portfolio theory (alas): Markowitz (1959).

The conventional probability paradigm: Most of the conventional discussions on probabilistic thought, especially in the philosophical literature, present minor variants of the same paradigm with the succession of the following historical contributions: Chevalier de Méré, Pascal, Cardano, De Moivre, Gauss, Bernouilli, Laplace, Bayes, von Mises, Carnap, Kolmogorov, Borel, De Finetti, Ramsey, etc. However, these concern the problems of *calculus* of probability, perhaps fraught with technical problems, but ones that are hair-splitting and, to be derogatory, *academic.* They are not of much concern in this book—because, in spite of my specialty, they do not seem to provide any remote usefulness for practical matters. For a review of these, I refer the reader to Gillies (2000), Von Plato (1994), Hacking (1990), or the more popular and immensely readable *Against the Gods* (Bernstein, 1996), itself drawing heavily on Florence Nightingale David (David, 1962). I recommend Bernstein's *Against the Gods* as a readable presentation of the history of probabilistic thought in engineering and the applied hard sciences but completely disagree with its message on the measurability of risks in the social sciences.

I repeat the point: To philosophers operating in probability *per se*, the problem seems one of calculus. In this book the problem of probability is largely a matter of knowledge, not one of computation. I consider these computations a mere footnote to the subject. The real problem is: Where do we get the probability from? How do we change our beliefs? I have been working on the "gambling with the wrong dice" problem: It is far more important to figure out what dice we are using when gambling than to develop sophisticated computations of outcomes and run the risk of having, say, dice with nothing but 6s. In economics, for instance, we have very large models of risk calculations sitting on very rickety assumptions (actually, not rickety but plain wrong). They smoke us with math, but everything else is wrong. Getting the right assumptions may matter more than having a sophisticated model.

An interesting problem is the "value at risk" issue where people imagine that they have a way to understand the risk using "complicated mathematics" and running predictions on rare events—thinking that they were able from past data to observe the probability distributions. The most interesting behavioral aspect is that those who advocate it do not seem to have tested their past predicting record, another Meehl type of problem.

Thinkers and philosophers of probability: Perhaps the most insightful book ever written on the subject remains the great John Maynard Keynes' *Treatise on Probability* (Keynes, 1989, 1920), which surprisingly has not collected dust—somehow everything we seem to discover appears to have been said in it (though, characteristic of Keynes, in a convoluted way). In the usual supplied lists of thinkers of probability, Shackle, who refined subjective probability, is often undeservedly absent (Shackle, 1973). Most authors also omit the relevant contributions of Isaac Levi on subjective probability and its links to belief (Levi, 1970), which should be required reading in that area (it is impenetrable but is worth the exercise). It is a shame because Isaac Levi is a probability *thinker* (as opposed to probability *calculator*). The epistemologist of probability Henry Kyburg (Kyburg, 1983) is also absent (too difficult to read).

One observation about philosophers as compared to scientists is that they do seem to work in a very heterogeneous and compartmented manner: Probability in philosophy is dealt with in different branches: logic, epistemology, rational choice, philosophy of mathematics, philosophy of science. It is surprising to see Nicholas Rescher delivering an insightful presidential address of the American Philosophical Association on the topic of luck (later published as a book called *Luck*, see Rescher, 1995) without discussing much of the problems in the philosophical and cognitive literature on probability.

Problems with my message: Note that many readers in the technical professions, say engineering, exhibited some difficulty seeing the connection between probability and belief and the importance of skepticism in risk management.

CHAPTER 14

Stoicism: Modern discussions in Becker (1998) and Banateanu (2001).

POSTSCRIPT

Uncertainty and pleasure: See Wilson, et. al. (2005) for the effect of randomness on the prolongation of positive hedonic states.

Looks and success: See (Shahami, et. al., 1993; Hosoda et. al., 1999). My friend Peter Bevelin wrote to me: "When I'm thinking about misjudgment of personalities I am always reminded of Sherlock Holmes in Arthur Conan Doyle's *The Sign of Four*. "It is of the first importance not to allow your judgment to be biased by personal qualities. I assure you that the most winning woman I ever knew was hanged for poisoning three little children for their insurance-money, and the most repellent man of my acquaintance is a philanthropist who has spent nearly a quarter of a million upon the London poor."

Maximizing: Psychology literature has focused on maximizing in terms of choice, not so much in these terms of actual optimization. I go beyond by looking at the activity of optimization in daily life. For a synthesis and review of the hedonic impact of maximizing and why "less is more," see Schwartz (2003). See also Schwartz, et. al. (2002). For the causal link between unhappiness and the pursuit of material benefits, see Kasser (2002).

Date of your death: I owe this last point to Gerd Gigerenzer.

Unpredictable behavior: See Miller (2000) for the discussion of the point in biology. See also Lucas's (1978) applications to a random monetary policy that thwarts expectations.

REFERENCES

Albouy, François-Xavier, 2002, *Le temps des catastrophes*. Paris: Descartes & Cie.

Al-Ghazali, 1989, "Mikhtarat Min Ahthar Al-Ghazali." In Saliba, Jamil, *Tarikh Al Falsafa Al Arabiah*. Beirut: Al Sharikah Al Ahlamiah Lilkitab.

Ambarish, R., and L. Siegel, 1996, "Time Is the Essence." *RISK*, 9, 8, 41–42.

Arnheim, Rudolf, 1971, *Entropy and Art: An Essay on Disorder and Order*. Berkeley: University of California Press.

Arrow, Kenneth, 1987, "Economic Theory and the Postulate of Rationality." In Eatwell, J., Milgate, M., and Newman, P., eds., 1987. *The New Palgrave: A Dictionary of Economics*, vol. 2, 69–74, London: Macmillan.

Arthur, Brian W., 1994, *Increasing Returns and Path Dependence in the Economy*. Ann Arbor: University of Michigan Press.

Banateanu, Anne, 2002, *La théorie stoïcienne de l'amitié: essai de reconstruction*. Fribourg: Editions Universitaires de Fribourg/Paris: Editions du Cerf.

Barabási, Albert-László, 2002, *Linked: The New Science of Networks*. Boston: Perseus Publishing.

Barber, B. M., and T. Odean, 2001, "The Internet and the Investor." *Journal of Economic Perspectives*, Winter, Vol. 15, No. 1, 41–54.

Barron, G., and I. Erev, 2003, "Small Feedback-based Decisions and Their Limited Correspondence to Description-based Decisions." *Journal of Behavioral Decision Making*, 16, 215–233.

Bates, Elisabeth, 1994, "Modularity, Domain Specificity, and the Development of Language." In Gajdusek, D.C., McKhann, G.M., and Bolis, C.L. eds., *Evolution and Neurology of Language: Discussions in Neuroscience*, 10(1–2), 136–149.

Bechara, A., A. R. Damasio, H. Damasio, and S. W. Anderson, 1994, "Insensitivity to Future Consequences Following Damage to Human Prefrontal Cortex." *Cognition*, 50:1–3, 7–15.

Becker, Lawrence C., 1998, *A New Stoicism*. Princeton, N.J.: Princeton University Press.

Bennett, Deborah J., 1998, *Randomness*. Cambridge, Mass.: Harvard University Press.

Bernstein, Peter L., 1996, *Against the Gods: The Remarkable Story of Risk*. New York: Wiley.

Berridge, Kent C., 2003, "Irrational Pursuits: Hyper-incentives from a Visceral Brain." In Brocas and Carillo.

Bouvier, Alban, ed., 1999, *Pareto aujourd'hui*. Paris: Presses Universitaires de France.

Brent, Joseph, 1993, *Charles Sanders Peirce: A Life*. Bloomington: Indiana University Press.

Brocas, I., and J. Carillo, eds., 2003, *The Psychology of Economic Decisions: Vol. 1: Rationality and Well-being*. Oxford: Oxford University Press.

Brock, W. A., and P.J.F. De Lima, 1995, "Nonlinear Time Series, Complexity Theory, and Finance." University of Wisconsin, Madison—Working Papers 9523.

Brock, W. A., D. A. Hsieh, and B. LeBaron, 1991, *Nonlinear Dynamics, Chaos, and Instability: Statistical Theory and Economic Evidence*, Cambridge, Mass.: MIT Press.

Brockman, John, 1995, *The Third Culture: Beyond the Scientific Revolution*. New York: Simon & Schuster.

Buchanan, Mark, 2002, *Ubiquity: Why Catastrophes Happen*. New York: Three Rivers Press.

Buehler, R., D. Griffin, and M. Ross, 2002, "Inside the Planning Fallacy: The Causes and Consequences of Optimistic Time Predictions." In Gilovich, Griffin and Kahneman.

Burnham, Terence C., 1997, *Essays on Genetic Evolution and Economics*. New York: Dissertation.com.

Burnham, Terence C., 2003, "Caveman Economics." Harvard Business School.

Burnham, T., and J. Phelan, 2000, *Mean Genes*. Boston: Perseus Publishing.

Camerer, C., G. Loewenstein, and D. Prelec, 2003, "Neuroeconomics: How Neuroscience Can Inform Economics. Caltech Working Paper.

Campbell, Jeremy, 1982, *Grammatical Man: Information, Entropy, Language and Life*. New York: Simon & Schuster.

Carter, Rita, 1999, *Mapping the Mind*. Berkeley: University of California Press.

Carter, Rita, 2002, *Exploring Consciousness*. Berkeley: University of California Press.

Chancellor, Edward, 1999, *Devil Take the Hindmost: A History of Financial Speculation*. New York: Farrar, Straus & Giroux.

Conlan, Roberta, ed., 1999, *States of Mind: New Discoveries About How Our Brains Make Us Who We Are*. New York: Wiley.

Cootner, Paul H., 1964, *The Random Character of Stock Market Prices*. Cambridge, Mass.: The MIT Press.

Cosmides, L., and J. Tooby, 1992, "Cognitive Adaptations for Social Exchange." In Barkow et al., eds., *The Adapted Mind*. Oxford: Oxford University Press.

Cover, T. M., and J. A. Thomas, 1991, *Elements of Information Theory*. New York: Wiley.

Csikszentmihalyi, Mihaly, 1993, *Flow: The Psychology of Optimal Experience*. New York: Perennial Press.

Csikszentmihalyi, Mihaly, 1998, *Finding Flow: The Psychology of Engagement with Everyday Life*. New York: Basic Books.

Damasio, Antonio, 1994, *Descartes' Error: Emotion, Reason, and the Human Brain*. New York: Avon Books.

Damasio, Antonio, 2000, *The Feeling of What Happens: Body and Emotion in the Making of Consciousness*. New York: Harvest Books.

Damasio, Antonio, 2003, *Looking for Spinoza: Joy, Sorrow and the Feeling Brain*. New York: Harcourt.

David, Florence Nightingale, 1962, *Games, Gods, and Gambling: A History of Probability and Statistical Ideas*. Oxford: Oxford University Press.

Dawes, R. M., D. Faust, and P. E. Meehl, 1989, "Clinical Versus Actuarial Judgment. *Science*, 243, 1668–1674.

Dawkins, Richard, 1989 (1976), *The Selfish Gene*. 2nd ed., Oxford: Oxford University Press.

De Vany, Arthur, 2003, *Hollywood Economics: Chaos in the Movie Industry*. London: Routledge.

Debreu, Gerard, 1959, *Theorie de la valeur,* Dunod, tr. *Theory of Value.* New York: Wiley.

Dennett, Daniel C., 1995, *Darwin's Dangerous Idea: Evolution and the Meanings of Life.* New York: Simon & Schuster.

Deutsch, David, 1997, *The Fabric of Reality.* New York: Penguin.

DeWitt, B. S., and N. Graham, eds., 1973, *The Many-Worlds Interpretation of Quantum Mechanics.* Princeton, N.J.: Princeton University Press.

Dugatkin, Lee Alan, 2001, *The Imitation Factor: Evolution Beyond the Gene.* New York: Simon & Schuster.

Easterly, William, 2001, *The Elusive Quest for Growth: Economists' Adventures and Misadventures in the Tropics.* Cambridge, Mass.: The MIT Press.

Edmonds, D., and J. Eidinow, 2001, *Wittgenstein's Poker: The Story of a Ten-Minute Argument Between Two Great Philosophers.* New York: Ecco.

Einstein, A., 1956 (1926), *Investigations on the Theory of the Brownian Movement.* New York: Dover.

Ekman, Paul, 1992, *Telling Lies: Clues to Deceit in the Marketplace, Politics and Marriage.* New York: W. W. Norton.

Elster, Jon, 1998, *Alchemies of the Mind: Rationality and the Emotions.* Cambridge, Eng.: Cambridge University Press.

Evans, Dylan, 2002, *Emotions: The Science of Sentiment.* Oxford: Oxford University Press.

Evans, D., and O. Zarate, 1999, *Introducing Evolutionary Psychology.* London: Totem Books.

Eysenck, M. W., and M. T. Keane, 2000, *Cognitive Psychology,* 4th ed.

Finucane, M. L., A. Alhakami, P. Slovic, and S. M. Johnson, 2000, "The Affect Heuristic in Judgments of Risks and Benefits." *Journal of Behavioral Decision Making,* 13, 1–17.

Fischhoff, Baruch, 1982, "For Those Condemned to Study the Past: Heuristics and Biases in Hindsight." In Kahneman, Slovic and Tversky.

Fodor, Jerry A., 1983, *The Modularity of Mind: An Essay on Faculty Psychology.* Cambridge, Mass.: The MIT Press.

Frank, Robert H., 1985, *Choosing the Right Pond: Human Behavior and the Quest for Status.* Oxford: Oxford University Press.

Frank, Robert H., 1999, *Luxury Fever: Why Money Fails to Satisfy in an Era of Excess.* Princeton, N.J.: Princeton University Press.

Frank, R. H., and P. J. Cook, 1995, *The Winner-Take-All Society: Why the Few at the Top Get So Much More Than the Rest of Us.* New York: Free Press.

Frederick, S., and G. Loewenstein, 1999, "Hedonic Adaptation," in Kahneman, Diener and Schwartz.

Freedman, D. A., and P. B. Stark, 2003, "What Is the Chance of an Earthquake?" Department of Statistics, University of California, Berkeley, CA 94720-3860. Technical Report 611. September 2001; revised January 2003.

Fukuyama, Francis, 1992, *The End of History and the Last Man.* New York: Free Press.

Galbraith, John Kenneth, 1997, *The Great Crash 1929.* New York: Mariner Books.

Gehring, W. J., and A. R. Willoughby, 2002, "The Medial Frontal Cortex and the Rapid Processing of Monetary Gains and Losses." *Science,* 295, March.

Georgescu-Roegen, Nicholas, 1971, *The Entropy Law and the Economic Process.* Cambridge, Mass.: Harvard University Press.

Gigerenzer, Gerd, 1989, *The Empire of Chance: How Probability Changed Science and Everyday Life.* Cambridge, Eng.: Cambridge University Press.

Gigerenzer, Gerd, 1996, "On Narrow Norms and Vague Heuristics: A Reply to Kahneman and Tversky. *Psychological Review,* 103, 592–596.

Gigerenzer, Gerd, 2003, *Calculated Risks: How to Know When Numbers Deceive You.* New York: Simon & Schuster.

Gigerenzer G., P. M. Todd, and ABC Research Group, 2000, *Simple Heuristics That Make Us Smart.* Oxford: Oxford University Press.

Gigerenzer, G., J. Czerlinski, and L. Martignon, 2002, "How Good Are Fast and Frugal Heuristics?" In Gilovich, Griffin, and Kahneman.

Gilbert, D., E. Pinel, T. D. Wilson, S. Blumberg, and T. Weatley, 2002, "Durability Bias in Affective Forecasting." In Gilovich, Griffin, and Kahneman.

Gillies, Donald, 2000, *Philosophical Theories of Probability.* London: Routledge.

Gilovich, T., R. P. Vallone, and A. Tversky, 1985, "The Hot Hand in Basketball: On the Misperception of Random Sequences." *Cognitive Psychology,* 17, 295–314.

Gilovich, T., D. Griffin, and D. Kahneman, eds., 2002, *Heuristics and Biases: The Psychology of Intuitive Judgment.* Cambridge, Eng.: Cambridge University Press.

Gladwell, Malcolm, 1996, "The Tipping Point: Why Is the City Suddenly

So Much Safer—Could It Be That Crime Really Is an Epidemic?" *The New Yorker*, June 3.

Gladwell, Malcolm, 2000, *The Tipping Point: How Little Things Can Make a Big Difference*. New York: Little, Brown.

———, 2002, "Blowing Up: How Nassim Taleb Turned the Inevitability of Disaster into an Investment Strategy." *The New Yorker*, April 22 and 29.

Glimcher, Paul, 2002, *Decisions, Uncertainty, and the Brain: The Science of Neuroeconomics*. Cambridge, Mass.: The MIT Press.

Goleman, Daniel, 1995, *Emotional Intelligence: Why It Could Matter More Than IQ*. New York: Bantam Books.

Goleman, Daniel, 2003, *Destructive Emotions, How Can We Overcome Them?: A Scientific Dialogue with the Dalai Lama*. New York: Bantam.

Goodman, Nelson, 1954, *Facts, Fiction and Forecast*. Cambridge, Mass.: Harvard University Press.

Hacking, Ian, 1990, *The Taming of Chance*. Cambridge, Eng.: Cambridge University Press.

Hacohen, Malachi Haim, 2001, *Karl Popper, The Formative Years, 1902–1945: Politics and Philosophy in Interwar Vienna*. Cambridge, Eng.: Cambridge University Press.

Hayek, F. A., 1945, "The Use of Knowledge in Society." *American Economic Review*, 35(4), 519–530.

Hayek, F. A., 1994, *The Road to Serfdom*. Chicago: University of Chicago Press.

Hilton, Denis, 2003, "Psychology and the Financial Markets: Applications to Understanding and Remedying Irrational Decision-making." In Brocas and Carillo.

Hirshleifer, J., and J. G. Riley, 1992, *The Analytics of Uncertainty and Information*. Cambridge, Eng.: Cambridge University Press.

Horrobin, David, 2002, *Madness of Adam and Eve: How Schizophrenia Shaped Humanity*. New York: Transworld Publishers Limited.

Hosoda, M., G. Coats, E. F. Stone-Romero, and C. A. Backus, 1999, "Who Will Fare Better in Employment-Related Decisions? A Meta-Analytic Review of Physical Attractiveness Research in Work Settings." Paper presented at the meeting of the Society of Industrial Organizational Psychology, Atlanta, Georgia.

Hsee, C. K., and Y. R. Rottenstreich, 2004, "Music, Pandas and Muggers: On the Affective Psychology of Value." Forthcoming, *Journal of Experimental Psychology*.

Hsieh, David A., 1991, "Chaos and Nonlinear Dynamics: Application to Financial Markets." *The Journal of Finance*, 46(5), 1839–1877.

Huang, C. F., and R. H. Litzenberger, 1988, *Foundations for Financial Economics*. New York/Amsterdam/London: North-Holland.

Hume, David, 1999 (1748), *An Enquiry Concerning Human Understanding*. Oxford: Oxford University Press.

Ingersoll, Jonathan E., Jr., 1987, *The Theory of Financial Decision Making*. Lanham, Md.: Rowman & Littlefield Publishing.

Jaynes, E. T., 2003, *Probability Theory: The Logic of Science*. Cambridge, Eng.: Cambridge University Press.

Kahneman, D., 2003, "Why People Take Risks." In *Gestire la vulnerabilità e l'incertezza: un incontro internazionale fra studiosi e capi di impresa*. Rome: Italian Institute of Risk Studies.

———, E. Diener, and N. Schwarz, eds., 1999, *Well-being: The Foundations of Hedonic Psychology*. New York: Russell Sage Foundation.

———, and S. Frederick, 2002, "Representativeness Revisited: Attribute Substitution in Intuitive Judgment." In Gilovich, Griffin, and Kahneman.

———, J. L. Knetsch, and R. H. Thaler, 1986, "Rational Choice and the Framing of Decisions." *Journal of Business*, Vol. 59 (4), 251–278.

———, J. L. Knetsch, and R. H. Thaler, 1991, "Anomalies: The Endowment Effect, Loss Aversion, and Status Quo Bias." In Kahneman and Tversky (2000).

———, and D. Lovallo, 1993, "Timid Choices and Bold Forecasts: A Cognitive Perspective on Risk-taking. *Management Science*, 39, 17–31.

———, P. Slovic, and A. Tversky, eds., 1982, *Judgment Under Uncertainty: Heuristics and Biases*. Cambridge, Eng.: Cambridge University Press.

———, and A. Tversky, 1972, "Subjective Probability: A Judgment of Representativeness." *Cognitive Psychology*, 3, 430–454.

———, and A. Tversky, 1973, "On the Psychology of Prediction." *Psychological Review*, 80: 237–251.

———, and A. Tversky, 1979, "Prospect Theory: An Analysis of Decision Under Risk." *Econometrica*, 47, 263–291.

———, and A. Tversky, 1982, "On the Study of Statistical Intuitions." *Cognition*, 11, 123–141.

———, and A. Tversky, 1996, "On the Reality of Cognitive Illusions." *Psychological Review*, 103, 582–591.

———, and A. Tversky, eds., 2000, *Choices, Values, and Frames*. Cambridge, Eng.: Cambridge University Press.

Kasser, Tim, 2002, *The High Price of Materialism*. Cambridge, Mass.: The MIT Press.

Keynes, John Maynard, 1937, "The General Theory." In *Quarterly Journal of Economics*, Vol. LI, 209–233.

———, 1989 (1920), *Treatise on Probability*. London: Macmillan.

Kindleberger, Charles P., 2001, *Manias, Panics, and Crashes*. New York: Wiley.

Knight, Frank, 1921 (1965), *Risk, Uncertainty and Profit*. New York: Harper and Row.

Kreps, David M., 1988, *Notes on the Theory of Choice*. Boulder, Colo.: Westview Press.

Kreps, J., and N. B. Davies, 1993, *An Introduction to Behavioral Ecology*, 3rd ed. Oxford: Blackwell Scientific Publications.

Kripke, Saul A., 1980, *Naming and Necessity*. Cambridge, Mass.: Harvard University Press.

Kurz, Mordecai, 1997, "Endogenous Uncertainty: A Unified View of Market Volatility," Working Paper. Stanford, Calif.: Stanford University Press.

Kyburg, Henry E., Jr., 1983, *Epistemology and Inference*. Minneapolis: University of Minnesota Press.

LeDoux, Joseph, 1998, *The Emotional Brain: The Mysterious Underpinnings of Emotional Life*. New York: Simon & Schuster.

LeDoux, Joseph, 2002, *Synaptic Self: How Our Brains Become Who We Are*. New York: Viking.

Levi, Isaac, 1970, *Gambling with Truth*. Boston, Mass.: The MIT Press.

Lewis, T., F. Amini, and R. Lannon, 2000, *A General Theory of Love*. New York: Vintage Books.

Lichtenstein, S., B. Fischhoff, and L. Phillips, 1977, "Calibration of Probabilities: The State of the Art." In Kahneman, Slovic, and Tversky (1982).

Loewenstein, G. F., E. U. Weber, C. K. Hsee, and E. S. Welch, 2001, "Risk As Feelings." *Psychological Bulletin*, 127, 267–286.

Lowenstein, Roger, 2000, *When Genius Failed: The Rise and Fall of Long-Term Capital Management*. New York: Random House.

Lucas, Robert E., 1978, "Asset Prices in an Exchange Economy." *Econometrica*, 46, 1429–1445.

Luce, R. D., and H. Raiffa, 1957, *Games and Decisions: Introduction and Critical Survey*. New York: Dover.

Machina, M. J., and M. Rothschild, 1987, "Risk." In Eatwell, J., Milgate, M., and Newman P., eds., 1987, *The New Palgrave: A Dictionary of Economics.* London: Macmillan.

MacKay, Charles, 2002, *Extraordinary Popular Delusions and the Madness of Crowds.* New York: Metro Books.

Magee, Bryan, 1997, *Confessions of a Philosopher.* London: Weidenfeld & Nicholson.

Mandelbrot, Benoit B., 1997, *Fractals and Scaling in Finance.* New York: Springer-Verlag.

Markowitz, Harry, 1959, *Portfolio Selection: Efficient Diversification of Investments,* 2nd ed. New York: Wiley.

Meehl, Paul E., 1954, *Clinical Versus Statistical Predictions: A Theoretical Analysis and Revision of the Literature.* Minneapolis: University of Minnesota Press.

Menand, Louis, 2001, *The Metaphysical Club: A Story of Ideas in America.* New York: Farrar, Straus & Giroux.

Merton, Robert C., 1992, *Continuous-Time Finance,* 2nd ed. Cambridge, Eng.: Blackwell.

Miller, Geoffrey F., 2000, *The Mating Mind: How Sexual Choice Shaped the Evolution of Human Nature.* New York: Doubleday.

Mumford, David, 1999, "The Dawning of the Age of Stochasticity." www.dam.brown.edu/people/mumford/Papers/Dawning.ps.

Myers, David G., 2002, *Intuition: Its Powers and Perils.* New Haven, Conn.: Yale University Press.

Nadeau, Maurice, 1970, *Histoire du surréalisme.* Paris: Seuil.

Niederhoffer, Victor, 1997, *The Education of a Speculator.* New York: Wiley.

Nozick, Robert, 1993, *The Nature of Rationality.* Princeton, N.J.: Princeton University Press.

Paulos, John Allen, 1988, *Innumeracy.* New York: Hill and Wang, a division of Farrar, Straus, and Giroux.

——, 2003, *A Mathematician Plays the Stock Market.* Boston: Basic Books.

Peirce, Charles S., 1998 (1923), *Chance, Love and Logic: Philosophical Essays.* Lincoln: University of Nebraska Press.

Peterson, Ivars, 1998, *The Jungles of Randomness: A Mathematical Safari.* New York: Wiley.

Piattelli-Palmarini, Massimo, 1994, *Inevitable Illusions: How Mistakes of Reason Rule Our Minds.* New York: Wiley.

Pinker, Steven, 1997, *How the Mind Works*. New York: W. W. Norton.

Pinker, Steven, 2002, *The Blank Slate: The Modern Denial of Human Nature*. New York: Viking.

Plotkin, Henry, 1998, *Evolution in Mind: An Introduction to Evolutionary Psychology*. Cambridge, Mass.: Harvard University Press.

Popper, Karl R., 1971, *The Open Society and Its Enemies*, 5th ed. Princeton, N.J.: Princeton University Press.

————, 1992, *Conjectures and Refutations: The Growth of Scientific Knowledge*, 5th ed. London: Routledge.

————, 1994, *The Myth of the Framework*. London: Routledge.

————, 2002, *The Logic of Scientific Discovery*, 15th ed. London: Routledge.

————, 2002, *The Poverty of Historicism*. London: Routledge.

Posner, Richard A., 2002, *Public Intellectuals: A Study in Decline*. Cambridge, Mass.: Harvard University Press.

Rabin, Mathew, 2000, "Inference by Believers in the Law of Small Numbers." Economics Department, University of California, Berkeley, Working Paper E00-282, http://repositories.cdlib.org/iber/econ/E00-282.

Rabin, M., and R. H. Thaler, 2001, "Anomalies: Risk Aversion." *Journal of Economic Perspectives*, 15(1), Winter, 219–232.

Ramachandran, V. S., and S. Blakeslee, 1998, *Phantoms in the Brain*. New York: Morrow.

Ratey, John J., 2001, *A User's Guide to the Brain: Perception, Attention and the Four Theaters of the Brain*. New York: Pantheon.

Rescher, Nicholas, 1995, *Luck: The Brilliant Randomness of Everyday Life*. New York: Farrar, Straus & Giroux.

Robbe-Grillet, Alain, 1985, *Les gommes*. Paris: Editions de Minuit.

Rozan, Jean-Manuel, 1999, *Le fric*. Paris: Michel Lafon.

Sapolsky, Robert M., 1998, *Why Zebras Don't Get Ulcers: An Updated Guide to Stress, Stress-Related Diseases, and Coping*. New York: W. H. Freeman & Co.

Sapolsky, Robert M. (and Department of Neurology and Neurological Sciences, Stanford University School of Medicine), 2003, "Glucocorticoids and Hippocampal Atrophy in Neuropsychiatric Disorders." Stanford University.

Savage, Leonard J., 1972, *The Foundations of Statistics*. New York: Dover.

Schleifer, Andrei, 2000, *Inefficient Markets: An Introduction to Behavioral Finance*. Oxford: Oxford University Press.

Schwartz, Barry, 2003, *The Paradox Of Choice.* New York: Ecco.

Schwartz, B., A. Ward, J. Monterosso, S. Lyubomirsky, K. White, and D. R. Lehman, 2002, "Maximizing Versus Satisficing: Happiness Is a Matter of Choice," *J Pers Soc Psychol.* Nov., 83 (5):1178–1197.

Searle, John, J., 2001, *Rationality in Action.* Cambridge, Mass.: The MIT Press.

Sen, Amartya, K., 1977, "Rational: A Critique of the Behavioral Foundations of Economic Theory. *Philosophy and Public Affairs,* 6, 317–344.

———, 2003, *Rationality and Freedom.* Cambridge, Mass.: The Belknap Press of Harvard University.

Shackle, George L. S., 1973, *Epistemics and Economics: A Critique of Economic Doctrines.* Cambridge, Eng.: Cambridge University Press.

Shahani, C., R. L. Dipboye, and T. M. Gehrlein, 1993, "Attractiveness Bias in the Interview: Exploring the Boundaries of an Effect." *Basic and Applied Social Psychology,* 14 (3), 317–328.

Shefrin, Hersh, 2000, *Beyond Fear and Greed: Understanding Behavioral Finance and the Psychology of Investing.* New York: Oxford University Press.

Shiller, Robert J., 1981, "Do Stock Prices Move Too Much to Be Justified by Subsequent Changes in Dividends?" *American Economic Review,* Vol. 71, 3, 421–436.

———, 1989, *Market Volatility.* Cambridge, Mass.: The MIT Press.

———, 1990. "Market Volatility and Investor Behavior." *American Economic Review,* Vol. 80, 2, 58–62.

———, 2000, *Irrational Exuberance.* Princeton, N.J.: Princeton University Press.

Shizgal, Peter, 1999, "On the Neural Computation of Utility: Implications from Studies of Brain Simulation Rewards." In Kahneman, Diener and Schwarz.

Sigelman, C. K., D. B. Thomas, L. Sigelman, and F. D. Ribich, 1986, "Gender, Physical Attractiveness, and Electability: An Experimental Investigation of Voter Biases." *Journal of Applied Social Psychology,* 16 (3), 229–248.

Simon, Herbert A., 1955, "A Behavioral Model of Rational Choice." *Quarterly Journal of Economics,* 69, 99–118.

———, 1956, "Rational Choice and the Structure of the Environment." *Psychological Review,* 63, 129–138.

———, 1957, *Models of Man.* New York: Wiley.

————, 1983, *Reason in Human Affairs*. Stanford, Calif.: Stanford University Press.

————, 1987, "Behavioral Economics." In Eatwell, J., Milgate, M., and Newman, P., eds., 1987, *The New Palgrave: A Dictionary of Economics*. London: Macmillan.

————, 1987, "Bounded Rationality." In Eatwell, J., Milgate, M., and Newman, P., eds., 1987, *The New Palgrave: A Dictionary of Economics*. London: Macmillan.

Skinner, B. F., 1948, "Superstition in the Pigeon." *Journal of Experimental Psychology*, 38, 168–172.

Sloman, Steven A., 1996, "The Empirical Case for Two Systems of Reasoning." *Psychological Bulletin*, 119, 3–22.

Sloman, Steven A., 2002, "Two Systems of Reasoning." In Gilovich, Griffin, and Kahneman.

Slovic, Paul, 1987, "Perception of Risk." *Science*, 236, 280–285.

————, 2000, *The Perception of Risk*. London: Earthscan Publications.

————, M. Finucane, E. Peters, and D. G. MacGregor, 2002, "The Affect Heuristic." In Gilovich, Griffin and Kahneman.

————, M. Finucane, E. Peters, and D. G. MacGregor, 2003, "Rational Actors or Rational Fools? Implications of the Affect Heuristic for Behavioral Economics." Working Paper. www.decisionresearch.com.

————, M. Finucane, E. Peters, and D. G. MacGregor, 2003, "Risk As Analysis, Risk As Feelings: Some Thoughts About Affect, Reason, Risk, and Rationality." Paper presented at the Annual Meeting of the Society for Risk Analysis, New Orleans, La., December 10, 2002.

Sokal, Alan D., 1996, "Transgressing the Boundaries: Toward a Transformative Hermeneutics of Quantum Gravity." *Social Text*, 46/47, 217–252.

Sornette, Didier, 2003, *Why Stock Markets Crash: Critical Events in Complex Financial Systems*. Princeton, N.J.: Princeton University Press.

Soros, George, 1988, *The Alchemy of Finance: Reading the Mind of the Market*. New York: Simon & Schuster.

Sowell, Thomas, 1987, *A Conflict of Visions: Ideological Origins of Political Struggles*. New York: Morrow.

Spencer, B. A., and G. S. Taylor, 1988, "Effects of Facial Attractiveness and Gender on Causal Attributions of Managerial Performance." *Sex Roles*, 19 (5/6), 273–285.

Stanley, T. J., 2000, *The Millionaire Mind*. Kansas City: Andrews McMeel Publishing.

————, and W. D. Danko, 1996, *The Millionaire Next Door: The Surprising Secrets of America's Wealthy*. Atlanta: Longstreet Press.

Stanovich, K., and R. West, 2000, "Individual Differences in Reasoning: Implications for the Rationality Debate." *Behavioral and Brain Sciences*, 23, 645–665.

Sterelny, Kim, 2001, *Dawkins vs Gould: Survival of the Fittest*. Cambridge, Eng.: Totem Books.

Stigler, Stephen M., 1986, *The History of Statistics: The Measurement of Uncertainty Before 1900*. Cambridge, Mass.: The Belknap Press of Harvard University.

————, 2002, *Statistics on the Table: The History of Statistical Concepts and Methods*. Cambridge, Mass.: Harvard University Press.

Sullivan, R., A. Timmermann, and H. White, 1999, "Data-snooping, Technical Trading Rule Performance and the Bootstrap." *Journal of Finance*, October, 54, 1647–1692.

Taleb, Nassim Nicholas, 1997, *Dynamic Hedging: Managing Vanilla and Exotic Options*. New York: Wiley.

————, 2004, "Bleed or Blowup? Why Do We Prefer Asymmetric Payoffs?" *Journal of Behavioral Finance*, 5.

Taszka, T., and P. Zielonka, 2002, "Expert Judgments: Financial Analysts Versus Weather Forecasters." *The Journal of Psychology and Financial Markets*, Vol 3(3), 152–160.

Thaler, Richard H., 1980, "Towards a Positive Theory of Consumer Choice," *Journal of Economic Behavior and Organization*, 1, 39–60.

————, 1994, *Quasi Rational Economics*. New York: Russell Sage Foundation.

————, 1994, *The Winner's Curse: Paradoxes and Anomalies of Economic Life*. Princeton, N.J.: Princeton University Press.

Toulmin, Stephen, 1990, *Cosmopolis: The Hidden Agenda of Modernity*. New York: Free Press.

Tversky, A., and D. Kahneman, 1971, "Belief in the Law of Small Numbers." *Psychology Bulletin*, Aug. 76(2), 105–110.

————, and D. Kahneman, 1973, "Availability: A Heuristic for Judging Frequency and Probability." *Cognitive Psychology*, 5, 207–232.

————, and D. Kahneman, 1982, "Evidential Impact of Base-Rates." In Kahneman, Slovic, and Tversky, 153–160.

————, and D. Kahneman, 1992, "Advances in Prospect Theory: Cumulative Representation of Uncertainty. *Journal of Risk and Uncertainty*, 5, 297–323.

Voit, Johannes, 2001, *The Statistical Mechanics of Financial Markets*. Heidelberg: Springer.

Von Mises, Richard, 1957 (1928), *Probability, Statistics, and Truth*. New York: Dover.

Von Plato, Jan, 1994, *Creating Modern Probability*. Cambridge, Eng.: Cambridge University Press.

Watts, Duncan, 2003, *Six Degrees: The Science of a Connected Age*. New York: W. W. Norton.

Wegner, Daniel M., 2002, *The Illusion of Conscious Will*. Cambridge, Mass.: The MIT Press.

Weinberg, Steven, 2001, *Facing Up: Science and Its Cultural Adversaries*. Working Paper. Harvard University.

Wilson, Timothy D., 2002, *Strangers to Ourselves: Discovering the Adaptive Unconscious*. Cambridge, Mass.: The Belknap Press of Harvard University.

Wilson, Edward O., 2000, *Sociobiology: The New Synthesis*. Cambridge, Mass.: Harvard University Press.

———, 2002, *The Future of Life*. New York: Knopf.

Wilson, T. D., D. B. Centerbar, D. A. Kermer, and D. T. Gilbert, 2005, "The Pleasures of Uncertainty: Prolonging Positive Moods in Ways People Do Not Anticipate," *J Pers Soc Psychol*. 2005 Jan.; 88 (1): 5–21.

———, D. Gilbert, and D. B. Centerbar, 2003, "Making Sense: The Causes of Emotional Evanescence." In Brocas and Carillo.

———, J. Meyers, and D. Gilbert, 2001, "Lessons from the Past: Do People Learn from Experience That Emotional Reactions Are Short Lived?" *Personality and Social Psychology Bulletin*.

Winston, Robert, 2002, *Human Instinct: How Our Primeval Impulses Shape Our Lives*. London: Bantam Press.

Zajdenweber, Daniel, 2000, *L'économie des extrèmes*. Paris: Flammarion.

Zajonc, R.B., 1980, "Feeling and Thinking: Preferences Need No Inferences." *American Psychologist*, 35, 151–175.

———, 1984, "On the Primacy of Affect." *American Psychologist*, 39, 117–123, 114.

Zizzo, D. J., and A. J. Oswald, 2001, "Are People Willing to Pay to Reduce Others' Incomes?" *Annales d'Economie et de Statistique*, July/December 63/64, 39–62.

INDEX

ABC (Adaptive Behavior and Cognition) Group, 200
Abelson, Alan, 62
absolute position, 16, 272
accumulators, 143–46
Achilles, 34
Adams, Evelyn, 160
adaptations, 201, 287
adverse selection, 158
affect heuristic, 191, 192, 196, 274, 285
Against the Gods, 289
Akerlof, George A., 123
Al-Ghazali, 236, 288
Albouy, François-Xavier, 274
Alexander the Great, 34
Alhakami, A., 274
alternative histories, 22, 23–24, 25, 45, 49–52
alternative medicine, 166–68
alternative sample paths, *see* sample paths
Ambarish, R., 281
Amini, F., 275, 286
amnesia, 53, 275
amygdala, 240, 288
anchoring, 191, 192, 193–94, 275
Arabic skeptics, 288
Aristotle, 271
Arnheim, Rudolf, 277

Arrow, Kenneth, 25, 283, 284
Arthur, Brian, 176, 281
astrology, 126, 127
asymmetry, 98, 105–6, 112–13, 130
attribution bias, 243, 244, 281
autokinetic effect, 277
availability heuristic, 195, 205, 274

Babbage, Nigel, 238
Bacchus, 247
backtesting, 162–64
Bacon, Francis, 117, 118
Banateanu, Anne, 290
Bank of Sweden Prize for Economics, 187
Barabási, Albert-László, 281
Barber, B. M., 288
Bates, Elisabeth, 286
Baudelaire, Charles, 77, 78
Bechara, A., 286
Becker, Lawrence C., 290
behavioral finance, 188–89, 287
bell curve, 99, 178, 282
Benjamin, Walter, 77
Bennett, Deborah, 206–7, 287
Bergson, Henri, 225
Bernstein, Peter L., 289
Berra, Yogi, 4, 35, 126
Berridge, Kent C., 285
Bevelin, Peter, 291

biases
 attribution, 243, 244, 281
 hindsight, 56, 191, 192, 269
 and randomness, 137–38
 selection, 158
 see also survivorship bias
The Bible Code, 160
birthday paradox, 159
black swan problem
 and Carlos, 159
 defined, 4, 26
 and induction, 117, 119, 126, 131
 and LCTM, 242
 and Merton's hedge fund, 62
 and Nero Tulip, 183–85
Blakeslee, S., 286
blindness, *see* option blindness;
 probability blindness
Bloomberg machine, 213–15
Boileau's adage, 39
bonds, emerging-market, 79–85
book reviews, 161–62, 223–24
book sales, 281–82
Borges, J. L., 77
borrowed wisdom, 39
Bouvier, Alban, 282
brain, *see* human brain
Brent, Joseph, 272
Breton, André, 76
Brock, W. A., 281
Brockman, John, 276
Buffett, Warren, 144
Bulhak, Andrew C., 73
bulls and bears, 50–51, 99–102
Burges, J. L., 77
Buridan, Jean, 180
Buridan's donkey, 180, 258
Burnham, Terence C., 198, 272, 285

cab drivers, New York City, 227
Caesar, Julius, 34, 174
call options, 208
Camerer, C., 287
camouflaged learning, 275
Campbell, Jeremy, 276
cancer clusters, 169–70
Canetti, Elias, 77
Carlos (emerging-market bonds
 trader), 79–85, 91, 159
Carnap, Rudolf, 71, 289
Carneades of Cyrene, 235–36
Carnegie-Mellon University, 187
Carter, Rita, 272, 286

catastrophes, 274
Cato the elder, 235–36
causality, 214–15, 284
cause-and-effect, 41–42
Cavafy, C. P., 64, 247–48
Centerbar, B., 275
certainty, 21, 25
 see also uncertainty
Chancellor, Edward, 275
chaos theory, 173–74
charlatanism, 115, 126, 160, 241
chess skills, 29–30
Chicago Board of Trade, 20
Chicago Mercantile Exchange, 5, 7
Cicero, 237
cigarette smoking, 233
civil servants, 55–56
Claparède, Edouard, 53, 203
Claudel, Paul, 125
Cleopatra, 174
clinical knowledge, 269–70
clusters, 169–70, 178–79
CNBC, 211–12, 225
CNN, 78
cognition, effect of emotions on, 274
cognitive errors in forecasting, 271
coin tossing, 152–53
coincidences, 157–61
compensation, corporate, 254–57
complexity, 36–37, 39, 177–78
Comte, Auguste, 128
conditional information, 223–24
conditional probabilities, 204, 207
Conlan, Roberta, 286
conspiracy theories, 161
contributions, visibility of, 254–57
conventional probability paradigm,
 289
corporate governance, 256
correctness and intelligibility, 39
correlation, 283
Cosmides, L., 285
courage, 271
Cover, T. M., 276
Credit Suisse First Boston, 226
criminal trials, 203–4
crisis hunters, 112
Croesus, King of Lydia, 3, 4, 57
cross-sectional problem, 86
Csikszentmihalyi, Mihaly, 288
currencies, 91–92, 108, 110, 261
Cyrus, King of Persia, 4
Czerlinski, J., 287

The Da Vinci Code, 161
Dada Engine, 73
daily newspapers, *see* journalism
Damasio, Antonio, 202–3, 273, 275, 286
Darwinism, 94–96, 277–78
data mining, 159, 160–61, 281
data snooping, 162
David, Florence Nightingale, 289
Davies, N. B., 285
Dawkins, Richard, 72, 95, 276, 284
De Lima, P. J. F., 281
De Vany, Art, 111–12, 279
Debreu, Gerard, 25, 177, 273
decision making under uncertainty, 60, 119
deductive statements, defined, 71
denial, 93
denigration of history, 26, 53, 54, 55
Derrida, Jacques, 72, 73
Descartes, R., 270
Descartes' Error, 202–3
Deutsch, David, 273
DeWitt, B. S., 273
Diaconis, Percy, 160
dialectics, 236
Diener, E., 276, 285
dip buyers, 82–83, 93
distilled thinking, 58, 59, 63
distribution, 156, 157–61
dog that did not bark, 170
domain-general adaptations, 286, 287
domain-specific adaptations, 201, 286, 287
Dos Passos, John, 162
Drosnin, Michael, 160
Dugatkin, Lee Alan, 277

ecological rationality, 200–201
econometrics, 107, 113–14
economics
    Hollywood, 279
    neoclassical, 123, 188–89
    as normative science, 189
    Soros' view, 122–23
    state space method, 25
    and uncertainty, 25, 188, 273
Edmonds, D., 280
efficient markets, 61–62, 189
Eidinow, J., 280
Einstein, Albert, 39–40, 48–49
elegance, personal, 249
Elster, Jon, 273

Eluard, Paul, 76
emerging-market bond trading, 79–85
emotional brain, 202, 203, 286
emotions
    effect on cognition, 274
    mad cow disease example, 38
    malnutrition example, 38
    and probability, 222–23
    and randomness, 68–69, 222–23
    and risk, 38–39, 273, 286–87
    social, 19, 273
    thinking with, 209
empiricism, 71, 117, 121, 188
empty suits, 255–56
endogenous uncertainty, 273
endowment effect, 239
entropy, 276–77
epiphenomenalism, 41–42, 274
epistemology, 49, 117
equity premium puzzle, 280
ergodicity, 57–58, 96, 156–57, 254
Evans, Dylan, 273, 285
Everett, Hugh, 25, 273
evolution
    misuse of Darwinism, 94–96
    naive theories, 94–96
    negative mutations, 95
    and Nozik, 106
    as probabilistic concept, 277–78
    proximate compared with ultimate cause, 284
    and randomness, 96
    survival of the fittest, 63
    theorists, 287
evolutionary biology, 50–51, 261
evolutionary psychology, 197–98, 200–201, 285–86
expectation of the maximum, 154
explicit memory, 275
"exquisite cadavers" exercise, 75–76, 276
eye contact, 223, 288
Eysenck, M. W., 270

fallibility, 272
falsificationism, 128
fame, 174–75
Fashionable Nonsense, 71–72
fear, and genetics, 274–75
feedback loops, 123, 277
Fertile Crescent, 199
Feynmann, Richard, 51
Finucane, M. L., 273, 274

firehouse effect, 85, 277
Fischhoff, Baruch, 269, 271
flow, 208, 288
Fodor, Jerry, 202, 285, 286
Frank, Robert, 272
Frederick, S., 285
frequency, 98–99, 105, 201
Friedman, Milton, 187, 272
Fukuyama, Francis, 274
funds of funds, 166

Galbraith, John Kenneth, 275
gamblers' ticks, 228–29
gambling, 98–99, 200, 241, 271, 277–78
"gambling with the wrong dice" problem, 289
Gates, Bill, 175–76
Gehring, W. J., 275
Georgescu-Roegen, Nicholas, 277
gerontocracy, 63–64
Gigerenzer, Gerd, 200, 287, 291
Gilbert, D., 275
Gillies, Donald, 289
Gilovich, T., 280, 285
Gladwell, Malcolm, 178–79, 282
Glincher, Paul, 287
global-warming debate, 105
"The God Abandons Antony," 247–48
Goleman, Daniel, 273, 275, 286
Goodman, Nelson, 279–80
Gorgias, 236
Gould, Steven Jay, 94–95, 97–98
Graham, N., 273
Grant, Jim, 62
Grasso, Richard, 256
Greene, Graham, 24, 225
Greenspan, Alan, 156
Griffin, D., 285

Hacking, Ian, 289
Hacohen, Malachi Haim, 280
happiness, uncertainty and, 257–61
Harsanyi, John, 123, 284
Hayek, F. A., 128, 280
hedge funds, 62, 83, 109, 110–11, 281
hedonic treadmill, 287
Hegel, George W. F., 74–75, 124, 276
heroes, 34, 144, 198–99, 246
heuristics, 188, 191–96, 197, 285, 287
Hilton, Denis, 271, 288
hindsight bias, 56, 191, 192, 269
Hirshleifer, J., 273

historical determinism, 55–56
history
  denigrating, 26, 53, 54, 55
  and financial markets, 57–58
  and journalism, 58, 67–68
  learning from, 51–52, 53–54
Hollywood economics, 279
Holmes, Sherlock, 170, 291
Homer, 198–99, 221
Horobin, David, 272
Hosoda, M., 291
hot-hand effect, 155, 280
Hsee, C. K., 273, 279
Hsieh, David A., 281
Huang, C. F., 273
human brain
  asymmetry, 275
  and linear combinations, 183
  and nonlinearities, 179
  physical and emotional, 201–2, 203, 286–87
  Triune theory, 201–2
Hume, David, 117, 236

Iliad, 34
illusion of knowledge, 288
imitation, 277
implicit memory, 275
In Search of Time Lost, 15, 238
induction
  defined, 130
  Goodman's riddle, 279–80
  and LTCM, 243
  and The Millionaire Next Door, 145–46
  Popper's answer, 126–28
  problem, 4, 26, 116–31
inductive statements, 71, 120
infallibilism, 272
inference
  affirming the consequent, 270
  as illusion of knowledge, 288
  inductive, 130
  and past performance, 135–37
  and problem of induction, 4, 26, 116–31
  statistical, 7, 127, 231, 271
information
  compared with noise, 58–61, 59, 64, 216, 217
  conditional, 223–24
  distilled compared with undistilled, 62

retired dentist example, 64–66, 67, 68
and risk, 38–39
as toxic, 60
Ingersoll, Jonathan E., 273
ingratitude factor, 26–27
intellectual curiosity, 27, 28, 81
intellectuals, scientific compared with literary, 70–75, 276
interest rate differential, 87, 88
International Monetary Fund, 85
introspection, 27, 28, 48, 60
investors
compared with traders, 92
ingratitude factor, 26–27
invisible histories, *see* alternative histories
Irrational Exuberance, 35
"It was obvious after the fact" effect, 191, 192

January effect, 157
Jaynes, E. T., 271
Jean-Patrice (boss), 31–33, 40, 55
John Doe A. (janitor), 21, 58
John Doe B. (dentist), 21
John (high-yield trader), 12–17, 86–90, 91, 96, 159
Johnson, S. M., 274
joint probability, 204–5
Jonas, Stan, 100
journalism
and emotion, 38–39
and George Will, 34–36
and history, 58, 67–68
noise compared with information, 58–61, 214, 215, 216, 217
and risk, 38–39
thoughtful compared with thoughtless, 62–63
and understanding, 210–18, 271

Kafka, Franz, 205
Kahneman, Daniel, 37, 52, 187–90, 195, 196, 197, 198, 271, 272, 275, 276, 278, 285, 287, 288
Kaletsky, Anatole, 62
Kant, Immanuel, 236
Kasser, Tim, 291
Keane, M. T., 270
Kennedy, Jacqueline, 247
Kenny (boss), 31, 32, 33, 40
keyboard, QWERTY, 175–76

Keynes, John Maynard, 48, 49, 127, 187, 188, 274, 277, 284, 290
Kindleberger, Charles P., 275
Knetsch, J. L., 272, 288
Knight, Frank, 188
knowledge, *see* epistemology
Kreps, David M., 285
Kripke, Saul, 25, 273
Krugman, Paul, 186
Kurz, Mordecai, 273
Kyburg, Henry, 290

Lannon, R., 275, 286
law of large numbers, 255
law of small numbers, 277
LeBaron, B., 281
LeDoux, Joseph, 202, 203, 274, 275, 286
Leibniz, Gottfried, 25, 95
les gommes, 274
leverage, 88, 89
Levi, Isaac, 290
Lewis, T., 275, 286
Lichtenstein, S., 271
life expectancy, 212–13
Linda problem, 196, 205
linear combinations, 183
Linnaeus, Carl, 95
literary intellectuals, 70–78
Litzenberger, R. H., 273
Loewenstein, G., 273, 287
long-term capital management, 275
Long Term Capital Management (LTCM) fund, 242–43
Los Alamos laboratory, 46
"loss of perspective" bias, 191
losses, stop, 84, 93, 131
Lourdes, France, 168
Lovallo, D., 271
Lowenstein, Roger, 275
LTCM (Long Term Capital Management) fund, 242–43
Lucas, Robert, 114, 279, 291
Luce, R. D., 284
luck
and coin tossing, 152–53
comparative, 165–66
and data snooping, 162
and "hot hand in basketball," 155
Machiavelli's view, 151
manufacturing with Monte Carlo engine, 152
and randomness, 150

Luck, 290
lucky fools, defined, 18

MacGregor, D. G., 274
Machiavelli, 151
MacKay, Charles, 275
mad cow disease, 38
Magee, Bryan, 129, 280
Mandelbrot, Benoit, 179, 282
Marc (New York lawyer), 139–43
market dips, buying, 82–83, 93
market volatility, 38–39, 61–62
Markowitz, Harry, 241–42, 288
Marr, David, 202
Marshall, Alfred, 187
Martignon, L., 287
Marxism, 73, 115
mathematicians
    compared with scientists, 28
    and Monte Carlo methods, 47
    pure, 28, 43–44, 47–48
mathematics
    compared with Monte Carlo
        simulations, 178
    and probability, 200, 234, 241, 271
MBAs, 29–30, 36–37, 262
McBirney, Peter, 270, 284
mean (average), 98, 106
media, *see* journalism
median, 98, 106
medicine
    alternative, 166–68
    and data mining, 164
    probabilities test, 206–7
    research and probabilities, 210–11
Meehl, Paul E., 244, 269–70
memory, 130, 275
Menand, Louis, 272
Merton, Robert C., 62, 242
Meyers, J., 275
Microsoft, 175–76
Milken, Michael, 146
Mill, John Stuart, 117
Miller, Geoffrey F., 278, 291
The Millionaire Next Door, 143–46, 270, 280
mistakes, 56
modularity, 285, 286
Monash University, 73, 276
"Monday morning quarterback"
    heuristic, 191, 192
Monod, Jacques, 94
Montaigne, 270

Monte Carlo simulations
    about, 44
    and alternative histories, 49–52
    building engines, 49–50
    coin tossing, 152–53
    compared with mathematics, 178
    creating alternative histories using, 45
    defined, 46
    and evolutionary biology, 50–51
    generating random runs, 46
    history, 46–47
    and literary discourse, 72, 73–74
    Polya process, 176–77
    retired dentist example, 64–66, 67, 68
Monte Carlos, *see* roulette wheels
Montherlant, Henry de, 245
Mosteller, Frederick, 160
Mumford, David, 271
Myers, David G., 274

Nadeau, Maurice, 276
naive Darwinism, 277–78
Nash, John, 123
negative skewness, 278–79
neoclassical economics, 123, 188–89
networks, 176, 178–79, 281
neurobiology, 201–3, 285, 288
New York cab drivers, 227
newspapers, *see* journalism
Newtonian physics, 126, 127
Niederhoffer, Victor, 117–21, 279
Nobel, Alfred, 114, 187, 241
Nobel Prize, 40, 167, 189, 241, 242
noise, 58–61, 62, 64–69, 214, 215, 216, 217, 232
nonlinear dynamics, 173, 281
nonlinearities, 172, 176, 179, 180–81
nonrandomness, 168–69
normative sciences, 189
Nozik, Robert, 106, 279

O. J. Simpson trial, 203–4, 224
O'Connell, Marty, 85
Octavius, 247
Odean, T., 288
Odyssey, 221
Omega TradeStation, 163
Onassis, Jacqueline Kennedy, 247
one-dimensional space, 206
optimizers compared with satisficers, 258–59
option blindness, 207–10

options, as asset, 207–10
Oswald, A. J., 272
outcomes
  compared with paths, 45
  as performance measures, 22
  possible, 21, 205–6, 234, 288
  and probabilities, 45
  realized compared with unrealized, 51
  underestimation, 288
outliers, 105, 155
overconfidence, 192, 279

Paglia, Camille, 73
parallel universes, 25, 71
Pareto-Levy distribution, 282
Pareto power laws, 178
Pascal, Blaise, 130, 173–74, 289
past performance
  comparing, 165–66
  counterintuitive properties, 151
  and inference, 135–37
  measures, 22
  misperceptions about, 151
path dependence, 175–76, 238, 239–40
Patrocles, 34
Pauling, Linus, 167
Pearson, Egon, 168
Pearson, Karl, 168–69
pecking order, 14, 16, 18, 102, 143, 272
Peirce, Charles S., 242, 272
Perse, Saint-John, 77
personal elegance, 249
Pesaran, Hashem, 114
peso problem, 109, 110
Peters, E., 274
Phelan, J., 285
Phillips, L., 271
philosopher bureaucrats, 185–87
philosophers
  possible worlds idea, 25
  and probability, 289, 290
  as scientists, 279
Philostratus, 64
physical brain, 201–2, 286
physics
  parallel universes, 25
  as positive science, 189
  as tool, 48
  as Wall Street trader background, 29–30

Pi, computing, 47
Piattilli-Palmarini, Massimo, 283
Pinker, Steven, 197, 272, 284, 285
Plato, 124
Plotkin, Henry, 277
Plutarch, 174
poetry
  and randomness, 75–78
  translated, 76–77
Polya process, 176–77
Popper, Karl, 71, 74, 119, 121–22, 124, 125, 126–28, 129, 177, 222, 231, 236, 237, 242, 276, 280
portfolio theory, 288
positive sciences, 189
positivism, 128
Posner, Richard A., 276
possible worlds, 25, 28, 273
power law dynamics, 178, 282
PowerPoint, 256
Powers, Jimmy, 287
Prelec, D., 287
probability
  and accountants, 27
  birthday paradox, 159
  conditional, 204, 207
  and criminal trials, 203–4
  defined, 21
  and deviation from the norm, 155
  difficulties understanding, 36–37
  and Einstein, 48–49
  and emotion, 222–23
  gambling example, 98–99
  as introspective, 48
  joint, 204–5
  and Keynes, 48, 49
  and language, 276–77
  and mathematics, 200, 234, 241, 271
  monkeys and typewriters, 135, 136–37
  Mumford's view, 271
  optimizing for, 200–201
  in philosophy, 289–90
  and Polya process, 177
  and Russian roulette, 23–24
  and skepticism, 33–34, 237
  and "small world" occurrences, 159–60
  stochastic processes, 45, 46
probability blindness, 283
proprietary trading, 8–12
Proust, Marcel, 15, 198, 238
proverbs and sayings, 39, 274

proximate cause, 284
pseudoscience, 115, 124
pseudoscientific historicism, 52, 274
pseudothinkers, 74–75
psychopaths, 19, 240, 272
public intellectuals, 276
publishing industry, 111
Pyrrho, 231

quants, 48, 195
quantum mechanics, 25, 28, 95
QWERTY keyboard, 175–76

Rabin, Yitzhak, 160
Raiffa, H., 284
Ramachandran, V. S., 286
random runs, 45–46
random sample paths, *see* random runs
random walk, 47, 49, 176
randomness
    in art and poetry, 75–78
    benefits of, 257–61
    and biases, 137–38
    and cancer clusters, 169–70
    compared with luck, 170–71
    compared with track records, 156
    and dynamics of fame, 174–75
    and emotion, 68–69, 222–23
    and evolution, 96
    fools of, 28, 42, 91–94
    and fund managers, 151
    and "funds of funds," 166
    and hedonic states, 291
    and "hot hand in basketball," 155
    and literary compared with scientific
        discourse, 72–78
    misperceptions about importance,
        151, 155
    and nonlinearity, 179–80
    and performance comparisons,
        165–66
    and probabilistic skepticism, 33–34
    and problem of induction, 116
    and regime switches, 95
    resistance to, 20, 27, 49
    scaling property, 66
    and Soros, 123
    and stress, 275
    what it looks like, 169
Randomness, 206–7
rare events, 4, 95, 96–97, 103–6,
    108–13, 243
Ratey, John J., 286

rational choice, 284
rationality, 283–84, 285
reality
    and false sense of security, 26
    generators, 40–41
    as viscious game, 26–27
reasoning systems, 196–97, 285
reference case problem, 169
regime switches, 63, 95
regression to the mean, 155
relative position, 16, 272
representativeness heuristic, 195
Rescher, Nicolas, 290
reverse survivors, 159
Riley, J. G., 273
risk
    and Alexander the Great, 34
    and alternative histories, 25
    aversion to, 9, 10, 12
    earthquake example, 37
    and emotions, 38–39, 273, 286–87
    impression of reduction, 41
    insurance policy example, 37
    and Jean-Patrice, 31–33
    and John, the high-yield trader,
        87–88
    and Julius Caesar, 34
    managing, 40–41
    in Russian roulette, 26
    underestimating, 210
    in viscious reality, 26
"risk as feelings" theory, 192, 273
risk avoidance, 38, 53, 273, 275,
    286–87
risk managers, 40–41
Robbe-Grillet, Alain, 274
Rogers, Jim, 104, 106
Rose, Lauren, 28
Rottenstreich, Y. R., 279
roulette wheels, 168–69
Rozan, Jean-Manuel, 239, 288
rules, 186–87, 188, 190–91
    see also heuristics; trading rules
Russian roulette, 23–24, 26

Sagan, Carl, 168
sample paths, 45, 58, 64–65, 96
Samuelson, Paul, 177, 187
sandpile effect, 172–74
Santa Fe Institute, 176, 178
Sapolsky, Robert M., 275
satisficing, 186, 187, 258–59
sayings and proverbs, 39, 274

scenario analysis, 25
Schadenfreude, 17, 250
schedules, 259–61
Schleifer, Andrei, 287
Scholes, Myron, 242
Schwartz, Barry, 291
Schwarz, N., 276, 285
scientific intellectuals, 70, 71, 197, 272, 276
scientism, 115, 284
scientists
    dual thinking, 30
    and Jean-Patrice, 33
    Russian, 29–30
    and science, 244
    street smarts, 30
    as Wall Street traders, 29–30
selection bias, 158
self-contradiction, 238–39
The Selfish Gene, 72, 276
Seneca, 248–49
sensationalism, 38
September 11 attacks, 39, 56
serotonin, 18–20, 272
Shackle, George L. S., 188, 290
Shahani, C., 291
Sharpe, William, 281
Sharpe ratios, 281
Shefrin, Hersh, 287
Shiller, Robert, 35–36, 61–62, 273, 275
Shizgal, Peter, 275
Siegel, L., 281
signals and noise, 216, 217, 232
    see also information
Simon, Herbert, 186–87, 189, 203, 285
skepticism, 33–34, 236, 237
skewness, 4, 98, 103, 177, 278–79
Skinner, B. F., 230–31, 288
Sloman, Steven A., 285
Slovic, P., 273, 274, 285
small numbers, 277, 278–79
"small world" occurrences, 159–60
Smith, Vernon, 189
smoothing kernel, 216
social emotions, 19, 273
sociobiology, 278, 285
Socrates, 235
Sokal, Alan, 71–72, 276
Solon, 4, 51–52, 56, 58, 96, 131
Sornette, Didier, 277, 281
Soros, George, 93, 104, 122–24, 125, 130, 144, 152, 239, 240, 243, 280
"sound-bite" effect, 191, 192

Soviet legal system, 190
Sowell, Thomas, 272
speeches, randomly constructing, 73–74
Spitznagel, Mark, 282
spontaneous remissions, 167–68
spurious survivors, 158
stalemates, 179–80
Stanley, T.J., 270, 280
Stanovich, K., 287
state space method, 25
states of nature, 25, 273
stationarity, 113
statistical arbitrageurs, 118
statistical inference, 7, 127, 231, 271
Stix, Gary, 62
stochastic processes, 45, 46
stoicism, 245–46, 248, 249, 290
stop losses, 84, 93, 131
stress, 40, 68, 275
Sullivan, R., 163, 281
superstition, 75, 228–29
survivorship bias
    and bad traders, 93
    and book reviews, 161–62
    and choice of profession, 21
    defined, 156
    and "funds of funds," 166
    and January effect, 157–58
    and Kenny, 31
    Marc and Janet example, 142–43
    and The Millionaire Next Door, 144–46
    mistake of ignoring, 146, 147–48
    in science, 170
    and size of initial population, 156–57
    and trading rules, 163
Sussman, Donald, 42
Swedish Central Bank Prize, 114, 187, 241

Taszka, T., 280
television
    financial experts on, 211–13
    weaning from, 224–25
Tempelsman, Maurice, 247
testable statements, 118–19, 120
Thaler, Richard H., 272, 287, 288
thinking, 185–87
Thomas, J. A., 276
time scale, retired dentist example, 64–66, 67, 68

time series analysis, 107, 108, 111, 113–14, 135, 151
Timmermann, A., 163, 281
The Tipping Point, 178, 282
Tooby, J., 285
Toulmin, Stephen, 270
track records, 136, 150–51, 154, 156, 157
traders
    aged, 63
    bulls *vs.* bears, 50–51
    compared with investors, 92
    compared with risk managers, 40
    option, 196–97, 209
    and probabilities, 196–97
    rational compared with irrational decisions, 232–33
    scientists as, 29–30
    those who last, 27–28
    traits of fools of randomness, 28, 91–94
    see also Carlos (emerging-market bonds trader); John (high-yield trader); Tulip, Nero
trading rules, 162, 163
Tranel, D., 286
translated poetry, 76–77
treadmill effect, 142, 259
Treasury bonds, 10, 11
Treatise on Laughter, 225
Treatise on Probability, 49, 274, 290
Tulip, Nero, 5–12, 20, 28, 104, 183–85, 250–51
Turing test, 72
Tversky, Amos, 37, 52, 187–90, 195, 196, 197, 198, 271, 275, 278, 280, 285, 287

ultimate cause, 284
Unabomber, 44
uncertainty
    and alternative histories, 25
    behavior under, 53, 183, 188, 278–79
    benefits of, 257–61
    decision making under, 60, 119
    and economics, 25, 188, 273
    endogenous, 273
    and happiness, 257–61, 291
    and knowledge, 235
unpredictability, 260, 261–62
U.S. dollar, 91–92, 108
utopia, 272

Vallone, R. P., 280
"value at risk" issue, 289
Vienna Circle, 70, 71, 77, 128
Voit, Johannes, 282
volatility, 61, 111, 112, 154
von Hayek, *see* Hayek, F. A.
von Mises, Richard, 289
Von Plato, Jan, 289

Walras, Leon, 177
Wason selection task, 285–86
Watts, Duncan, 281
We (novel), 284–85
wealth, resetting, 193, 275–76, 287
Weber, E. U., 273
Wegner, Daniel M., 274
Weinberg, Steven, 276
Welch, E. S., 273
West, R., 287
what-if analysis, 25
White, H., 163, 281
Will, George, 34–36, 41, 58, 62, 78
Willoughby, A. R., 275
Wilmott, Paul, 161–62
Wilson, E. O., 275, 278, 285, 291
Winston, Robert, 285
Wittgenstein, Ludwig, 71
Wittgenstein's ruler, 224, 242
work ethics, 11, 124

yield hogs, 83
    see also hedge funds

Zajdenweber, Daniel, 282
Zajonc, R. B., 285
Zamyatin, Yevgeny, 284–85
Zarate, O., 285
Zeno of Kition, 248
Zielonka, P., 280
Zizzo, D. J., 272
Zorglubs, 50–51

PENGUIN CLASSICS

THE WEALTH OF NATIONS
ADAM SMITH

Edited with an introduction and notes by Andrew Skinner

'It is not from the benevolence of the butcher, the brewer, or the baker that we expect our dinner, but from their regard to their own interest'

With this landmark treatise on political economy, Adam Smith paved the way for modern capitalism, arguing that a truly free market – fired by competition yet guided as if by an 'invisible hand' to ensure justice and equality – was the engine of a fair and productive society. *The Wealth of Nations* examines the 'division of labour as the key to economic growth, by ensuring the interdependence of individuals within society. Smith's work laid the foundations of economic theory in general and 'classical' economics in particular, but the real sophistication of his analysis derives from the fact that it also encompasses a combination of ethics, philosophy and history to create a vast panorama of society.

Published in two volumes (Books I-III and Books IV-V), this edition contains an in-depth discussion of Smith as an economist and social scientist, as well as a preface, further reading and explanatory notes.

PENGUIN POLITICS

GLOBALIZATION AND ITS DISCONTENTS
JOSEPH STIGLITZ

'A massively important political as well as economic document ... we should listen to him urgently' Will Hutton, *Guardian*

Our world is changing. Globalization is not working. It is hurting those it was meant to help. And now, the tide is turning ...

Explosive and shocking, *Globalization and Its Discontents* is the bestselling exposé of the all-powerful organizations that control our lives – from the man who has seen them at work first hand.

As Chief Economist at the World Bank, Nobel Prize-winner Joseph Stiglitz had a unique insider's view into the management of globalization. Now he speaks out against it: how the IMF and WTO preach fair trade yet impose crippling economic policies on developing nations; how free market 'shock therapy' made millions in East Asia and Russia worse off than they were before; and how the West has driven the global agenda to further its own financial interests.

Globalization *can* still be a force for good, Stiglitz argues. But the balance of power has to change. Here he offers real, tough solutions for the future.

'Compelling ... This book is everyone's guide to the misgovernment of globalization' J. K. Galbraith

'Stiglitz is a rare breed, an heretical economist who has ruffled the self-satisfied global establishment that once fed him. *Globalization and Its Discontents* declares war on the entire Washington financial and economic establishment' Ian Fraser, *Sunday Tribune*

'Gripping ... this landmark book ... shows him to be a worthy successor to Keynes' Robin Blackburn, *Independent*

PENGUIN POLITICS / ECONOMICS

THE TRUTH ABOUT MARKETS: WHY SOME NATIONS ARE RICH BUT MOST REMAIN POOR
JOHN KAY

'Readers of this illuminating book will better understand what has gone wrong with the market economy and what should be done about it' Joseph Stiglitz, author of *Globalization and Its Discontents*

'Ambitious and brilliantly executed ... accessible and witty ... John Kay exposes the flaws of the American business model' *The Times*

Capitalism faltered at the end of the 1990s as corporations were rocked by fraud and the state of the stock-market, and the American business model – unfettered self-interest, privatization and low tax – faced a storm of protest. But what are the alternatives to the mantras of market fundamentalism?

Leading economist John Kay unravels the truth about markets, from Wall Street to Switzerland, from Russia to Mumbai, examining why some nations are rich and some poor, why 'one-size-fits-all' globalization hurts developing countries and why markets *can* work – but only in a humane social and cultural context. His answers offer a radical new blueprint for the future.

'Profound ... a landmark work' Will Hutton

'Kay shows how markets really work – everything you wanted to know about economics but were afraid to ask' Mervyn King, Governor of the Bank of England

'Written with wit and subtlety ... An important contribution to the post-1990s reassessment of capitalism' Martin Vander Weyer, *Daily Telegraph*

'The big idea of the moment ... offers one of the most truthful and fruitful ways in years of looking at the relationship between modern government and the modern economy' Martin Kettle, *Guardian*

PENGUIN ECONOMICS

THE ECONOMICS OF INNOCENT FRAUD: TRUTH FOR OUR TIME
JOHN KENNETH GALBRAITH

'A prophet whose warnings have come to pass ... Galbraith is an iconoclast'
Independent

John Kenneth Galbraith, lifelong critic of unbridled corporate power and one of the most renowned economists of the twentieth century, delivers a scathing polemic on today's economics, politics and public morality.

Sounding the alarm on the gap between 'conventional wisdom' and reality, Galbraith distils years of expertise in this radical critique of our society. He shows the danger of the private sector's unprecedented and unbridled control over public life – from government to the military to the environment. And he reveals how politicians and the media have colluded in the myths of a benign 'market': that big business always knows best, that minimal intervention stimulates the economy, that obscene pay gaps and unrestrained self-enrichment are an inevitable by-product of the system. The result, he shows, is that we have given ourselves over to a lie and come to accept legal, legitimate, innocent fraud.

Galbraith's taut, wry and incisive analysis shows that the gulf between truth and illusion has never been wider. It is essential reading for anyone who cares about the economic and political future of the world.

John Kenneth Galbraith is Paul M. Warburg Professor of Economics, Emeritus, at Harvard University. He has worked in economics for over seventy years, and his many books include *A History of Economics*, *The Great Crash, 1929*, *The Age of Uncertainty* and *The Culture of Contentment*.

'The scourge of contemporary economics ... he has always been superb at attacking "conventional wisdom"' *Observer*

'America's leading public intellectual' *Guardian*

'The most widely read economist in the world' Amartya Sen, Nobel Prize-winner for Economics

PENGUIN HISTORY

THE CASH NEXUS : MONEY AND POWER IN THE MODERN WORLD 1700–2000
NIALL FERGUSON

Money : the root of all evil, or the stuff that makes the world go round?

Modern history shows that a nation's success largely depends on the way it manages its money. In times of war, finance has been just as crucial to victory as firepower. But where do money and politics meet? Starting in 1700 and ending at the present day, Niall Ferguson offers a bold and dazzling analysis of the evolution of today's economic and political landscape. Far from being driven by the profit motive alone, our recent history, as Ferguson makes brilliantly clear, has also been made by potent and often conflicting human impulses – sex, violence and the desire for power.

'A marvellous combination of persuasion and provocation...*The Cash Nexus* has enough ideas for a dozen books' Martin Daunton, *History Today*

'Niall Ferguson is probably the most brilliant of the up-and-coming generation of British historians...*The Cash Nexus* is...packed with intriguing arguments and controversial propositions...[an] outstanding book' Frank McLynn, *Independent*

'Combining scholarly research with trenchant analysis of contemporary issues, Ferguson explores the interaction between economics and politics from a variety of angles...he brings to these questions an encyclopaedic knowledge of economic history, and an array of unexpected anecdotes' Geoffrey Owen, *Sunday Telegraph*

'Ferguson is one of the most technically accomplished historians writing today...*The Cash Nexus* offers an important corrective to the naïve story of economic growth' Robert Skidelsky, *New York Review of Books*

He just wanted a decent book to read ...

Not too much to ask, is it? It was in 1935 when Allen Lane, Managing Director of Bodley Head Publishers, stood on a platform at Exeter railway station looking for something good to read on his journey back to London. His choice was limited to popular magazines and poor-quality paperbacks – the same choice faced every day by the vast majority of readers, few of whom could afford hardbacks. Lane's disappointment and subsequent anger at the range of books generally available led him to found a company – and change the world.

'We believed in the existence in this country of a vast reading public for intelligent books at a low price, and staked everything on it'
Sir Allen Lane, 1902–1970, founder of Penguin Books

The quality paperback had arrived – and not just in bookshops. Lane was adamant that his Penguins should appear in chain stores and tobacconists, and should cost no more than a packet of cigarettes.

Reading habits (and cigarette prices) have changed since 1935, but Penguin still believes in publishing the best books for everybody to enjoy. We still believe that good design costs no more than bad design, and we still believe that quality books published passionately and responsibly make the world a better place.

So wherever you see the little bird – whether it's on a piece of prize-winning literary fiction or a celebrity autobiography, political tour de force or historical masterpiece, a serial-killer thriller, reference book, world classic or a piece of pure escapism – you can bet that it represents the very best that the genre has to offer.

Whatever you like to read – trust Penguin.